Sedimentary Petrology - Implications in Petroleum Industry

Edited by Ali Ismail Al-Jubou ry

Published in London, United Kingdom

IntechOpen

Supporting open minds since 2005

Sedimentary Petrology - Implications in Petroleum Industry
http://dx.doi.org/10.5772/intechopen.92930
Edited by Ali Ismail Al-Juboury

Contributors
Ali Ismail Al-Juboury, Mohammed Al-Haj, Aboosh Al-Hadidy, Naveen Kumar, Naresh Kumar, Christian A. A. Paternina, Mai Hoang Dam, Lieu Kim Phuong, Nguyen Tan Trieu, Salvatore Critelli, Sara Criniti

First published in London, United Kingdom, 2022 by IntechOpen
IntechOpen is the global imprint of INTECHOPEN LIMITED, registered in England and Wales, registration number: 11086078, 5 Princes Gate Court, London, SW7 2QJ, United Kingdom
Printed in Croatia

British Library Cataloguing-in-Publication Data
A catalogue record for this book is available from the British Library

Additional hard and PDF copies can be obtained from orders@intechopen.com

Sedimentary Petrology - Implications in Petroleum Industry
Edited by Ali Ismail Al-Juboury
p. cm.
Print ISBN 978-1-83969-300-7
Online ISBN 978-1-83969-301-4
eBook (PDF) ISBN 978-1-83969-302-1

For EU product safety concerns:
IN TECH d.o.o., Prolaz Marije Krucifikse Kozulić 3, 51000 Rijeka, Croatia, info@intechopen.com or visit our website at intechopen.com.

Meet the editor

Prof. Dr. Ali Ismail Al-Juboury is a professor in the Geology Department, Mosul University, Iraq. He obtained his BSc in Geology and MSc in Sedimentology from Mosul University in 1980 and 1983, respectively, and his Ph.D. from Comenius University, Slovakia, in 1992. He has published 115 scientific papers (44 Clarivate and Scopus) in local and peer-reviewed journals in the fields of petroleum geology, sedimentology, geochemistry, and economic geology. He is a member of numerous international societies and serves on the editorial board of the *Iraqi Geological Journal*, *International Sedimentology, Stratigraphy Journal of Oil and Gas Basins*, and *International Journal of Geophysics and Geochemistry*. Dr. Al-Juboury has received several awards, including the Distinguished Scholars Award from the Arab Fund for Economic and Social Development, Kuwait, in 2009, and the Science and Technology (Geology) Award from the Islamic States Organization in 2014.

Contents

Preface

Sedimentary petrology is concerned with the description of sedimentary rocks, the processes of transportation and deposition of sedimentary materials, these materials' depositional environments, and the diagenetic processes acting on sediments after deposition. This science is of great importance in the petroleum industry since it concerns source rocks, occurrence and distribution of hydrocarbons, migration, and reservoir traps.

Sedimentary Petrology - Implications in Petroleum Industry includes five chapters that focus on the petrology of sedimentary rocks and their relation to provenance history, reservoir characterization, and hydrocarbon exploration.

Chapter 1, "Sandstone Petrology and Provenance in Fold Thrust Belt and Foreland Basin System" by Salvatore Critelli and Sara Criniti, summarizes the typical fold-thrust belt and foreland basin system settings from the petrological point of view of the key sandstone suites. This kind of reasoning highlights the evolutionary events of these tectonic frameworks.

Chapter 2, "Stylolite in Upper Cretaceous Carbonate Reservoirs from Northwestern Iraq" by Ali Al-Juboury, Mohammed A. Al-Haj and Aboosh H. Al-Hadidy, discusses the pressure solution features or stylolite that dominate in upper Cretaceous carbonate reservoirs from northwestern Iraq, which is regarded as the main reservoir in the upper Cretaceous succession in the region. Stylolites act as a barrier for vertical movement of hydrocarbons and prevent further movement of mineralization processes such as silicification of carbonate (calcite or dolomite).

Chapter 3, "Applications of Surfactants and Nanoparticles in Enhanced Oil Recovery Processes" by Christian A. Paternina, examines surfactant injection as a promising enhanced oil recovery method that reduces the interfacial tension of the water-oil interface to recover an additional amount of petroleum into the reservoir. The use of new techniques to reduce surfactant adsorption has been applied to increase the efficiency of the recovery factor. The most popular among nanoparticles have shown excellent performance because they reduce the adsorption of surfactant to interact with the rock surface as well as the surfactant.

Chapter 4, "Petrological and Biostratigraphic Characteristics of Pre-Cenozoic Carbonate Rocks in the Northern Song Hong Basin, Vietnam" by Mai Hoang Dam, Nguyen Tan Trieu and Lieu Kim Phuong, reveals the stratigraphic relationship that exists between the carbonate formations in Cat Ba Island and the basement rock in the northern Song Hong basin, Vietnam, and provides chronostratigraphic data that can be used in geological models for hydrocarbon exploration. Most of the carbonate rocks in the Carboniferous period are major dolostone and minor crystalline limestone, wackestone, and packstone.

Chapter 5, "Petro-Mineralogical and Geochemical Study of the Acid Magmatic Rocks of Tusham Ring Complex, NW Peninsular India" by Naveen Kumar and

Naresh Kumar, presents field and petrographical observations that are important for explaining the magmatic evolution and geodynamic setting of the Tusham Ring Complex (TRC), India.

This work was achieved with the great support of expert reviewers whose comments and contributions played an important and helpful role in securing the high quality of the included chapters. We express our gratitude to John S. Armstrong-Altrin, Dicle Bal Akkoca, Salim Hamed Hussain, Hamed Yarahmadzahi, Rahma S. Al-Auqadi, and Ibtisam Kamal.

We are also indebted to IntechOpen for publishing this book.

Dr. Ali Ismail Al-Jouboury
Professor,
Mosul University,
Mosul, Iraq

Chapter 1

Sandstone Petrology and Provenance in Fold Thrust Belt and Foreland Basin System

Salvatore Critelli and Sara Criniti

Abstract

The sandstone composition of foreland basin has a wide range of provenance signatures, reflecting the interplay between flexed underplate region and abrupt growth of the accreted upper plate region. The combination of contrasting detrital signatures reflects these dual plate interactions; indeed, several cases figure out that the earliest history of older foreland basin infilling is marked by quartz-rich sandstones, with cratonal or continental-block provenance of the flexed underplate flanks. As upper plate margin grows over the underplate, the nascent fold-and-thrust belt starts to be the main producer of grain particles, reflecting the space/time dependent progressive unroofing of the subjacent orogenic source terranes. The latter geodynamic processes are mainly reflected in the nature of sandstone compositions that become more lithic fragment-rich and feldspar-rich as the fold-thrust belt involves the progressive deepest portions of upper plate crustal terranes. In this context sandstone signatures reflect quartzolithic to quartzofeldspathic compositions.

Keywords: Sandstone, Petrostratigraphy, Convergent plate setting, Peripheral Foreland Basins, Sandstone Detrital Modes

1. Introduction

The evolutionary record of Earth's processes ascribed to sedimentary rocks has been pivotal in paleogeographical and paleotectonic reconstructions of source/basin systems. Compositional trends of clastic strata through space and time are used to frame the structural history of adjacent mountain belts and to monitor the key geodynamic changes during orogenic processes [1–5]. Clastic infilling of sedimentary basins, in orogenic systems, have been used as important indicators of tectonic activity and climatic changes. In the orogenic systems, clastic sedimentation may record the accretionary processes, the accommodation of the thrust units, and the flexural features of the foreland plate [5]. The chapter summarize the close relations between source to sink in orogenic-derived sandstones. Foreland basin stratigraphy is intimately connected with the growth of orogenic systems. The derived sandstones reflect the changing nature of fold-thrust belts and the flexural feautures of the underplate. The main goal of this contribution is to discuss the changing sandstone petrofacies during infill of foreland basin systems using a petrological approach.

IntechOpen

2. Fold-thrust belts in orogenic setting and Foreland Basin systems

The development of an orogenic wedge during continental collision points out a thickening of the crust. The excess of mass in this thickened crust acts like a load on the underthrust plate; that drag it to a flexure downwards close to the load, showing out a foreland basin framework [6, 7]. During plate convergence, the vertical load of the mountain belt migrates over the foreland plate, causing the migration of the associated foreland basin.

The foreland is the region between the front of a thrust belt and the adjacent craton [8–11]. Large volumes of clastic sediments derive from erosion of the thrust belt and subsequent deposition in the foreland basin. The foreland basin is generally defined as an elongate trough formed between a linear constructional orogenic belt and the stable craton, mainly in response to flexural subsidence caused by thrust-sheet loading in the orogen (**Figure 1**).

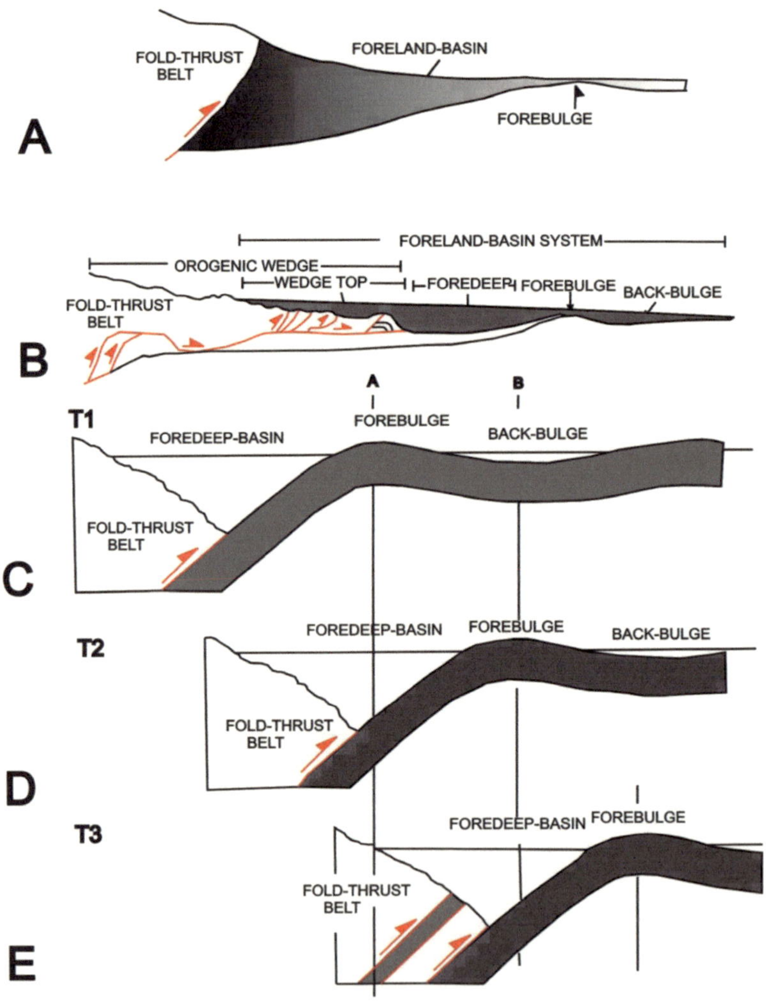

Figure 1.
Diagrammatic cross sections (from west to east) showing the general accepted notion of foreland-basin geometry (A, B) and the relationships between lithospheric flexure and accommodation space in foreland systems (C to E) [12–15]. A) General relationship between fold-thrust-belt, foreland basin and forebulge; B) foreland-basin geometry and depozones: Wedge-top, foredeep, forebulge and back-bulge depozones; C-E) relationship of the flexural features in times: c) initial (time 1) foreland system; D) foreland evolution during accretion of the fold-thrust-belt at time 2; forebulge migrated cratonward; E) previous forebulge assembled within the fold-thrust-belt. Modified after [4, 5, 12–15].

Foreland basin stratigraphy records tectonic, eustatic, and climatic changes at convergent plate margins [11], where the formation of unconformities is the result of the interplay of temporal variations in the erosion and lateral progradation rates of the orogenic wedge, as well as tectonic and eustatic sea-level changes [6, 16–18].

In foreland settings, subsidence and uplift are profoundly affected by lithospheric flexure. Foreland basin subsidence is primarily controlled by downflexing of the lithosphere in response to thrust accommodation and loading [6, 16, 19]. Subsidence rate gradually decreases away from the thrust front, producing an asymmetrical depression. Flexure uplift (forebulge) occurs as an isostatic response to warping downward and forms the distal margin of the foreland basin. Cratonward the forebulge flexure, a broad shallow downwarp or intrashelf basin forms, named as back-bulge basin (**Figure 1**) [12, 20].

The dimension and amount of flexural subsidence and uplift produced by the flexural features (i.e., foreland basin, forebulge, back-bulge basin) primarily depend on the geometry and density of the tectonic load, rheology of the lithosphere, density and volume of the sediment infill, and amount of thrust wedge and forebulge erosion [5, 6, 12, 16, 21]. The interrelationships between lithospheric flexure, single thrust accommodation within the accretionary wedge structure and flexural subsidence creates geometrical complex bodies within the foreland region. The foreland basin system may be divided into four depozones, the wedge-top, the foredeep, the forebulge, and the back-bulge depozones [12] (**Figure 1**). Boundary between depozones may shift laterally through time following the deformation propagation. The longitudinal dimension of the foreland basin system is roughly equal to the length of the adjacent fold-thrust belt [12].

3. Concepts of compositional signatures

The petrographic signatures during the different ages of the foreland infill show a clear sign of the diverse provenance sands, related to the different steps of the uplift stages and strictly connected to nearby rock portions exposed during the tectonic events. The discrimination of the principal source than the secondary one will point out the whole petrographic composition. This approach gives the base to the building of provenance and regional scale models from sandstone petrography.

3.1 Provenance models

Foreland regions are one of the typical setting in which huge volumes of clastic sediments are rapidly accumulated. This peculiar feature became a key element in provenance studies of such tectonic setting, because it consequently reveals several issues as the framework of the basin evolution complex history, sediment dispersal pathways, dating of major thrust events and the thrust-belt unroofing history [2, 22–25]. In this kind of setting, the uplift-erosion-transport-deposition system is genetically and intimately related to the deformation style in thin-skinned thrusted terranes. Transport of clastic sediment parallel to the tectonic shifting is the commonly assumed setting for the clastic-wedge/thrust association [23–25], named as «synthetic dispersal». However, opposite sediment dispersal pathways to the tectonic shifting is possible where hanging-wall beds dip toward the interior of the thrust belt, paleoslope, and therefore sediment dispersal, is opposite to tectonic transport. Such dispersal model, defined as «anthitetic dispersal» [25], reveals a kind of inverted stratigraphy, reflecting unroofing in the source, or mixed compositions. As result of these assumptions, it follows that sediment dispersal pathways in foreland basin systems are controlled by geometries within the thrust sheet system as frontal ramps, lateral ramps, and diverse hanging-wall beds dip. If distinct source-rock compositions are eroded sequentially, as in the case of predominantly vertical uplift of a stratigraphic section, «unroofing sequences» are commonly formed in the resultant anatomy of a clastic wedge [23, 25, 26].

This erosional inverted clast stratigraphy (unroofing history) can provide valuable information about the evolving sources and the identification of specific source areas [26–30]. In the case of thin-skinned thrusted terrains (where horizontal transport dominates), layered rocks with different lithologies are exposed to erosion as they pass over a ramp, providing a blended clastic dispersal pathways of the exposed rock types. The resulting clastics may show no unroofing sequences, but include the same blended clast composition for relatively great thickness. These blended clastics may indicate that the source rocks were formed by tectonic transport over a ramp [25]. In thin-skinned thrust belts, both «unroofing sequences» and «blended clastics» can result in combinations, due to the evolution of the entire orogenic system and foreland basin infill.

3.2 Large and regional-scale models based on sandstone Petrostratigraphy

Numerous studies have demonstrated that sand (stone) from foreland basins are characterized by high framework percentages of quartz and unstable sedimentary and metamorphic lithic fragments, and the mean composition is quartzolithic [1–3, 5, 30–32].

The foreland basin systems are a typical basin-setting in which multiple sources can be active at the same time, and the derivative sandstones may show mixed petrofacies [4, 5, 31, 33] (**Figure 2**). Schwab [31], in a general statement of

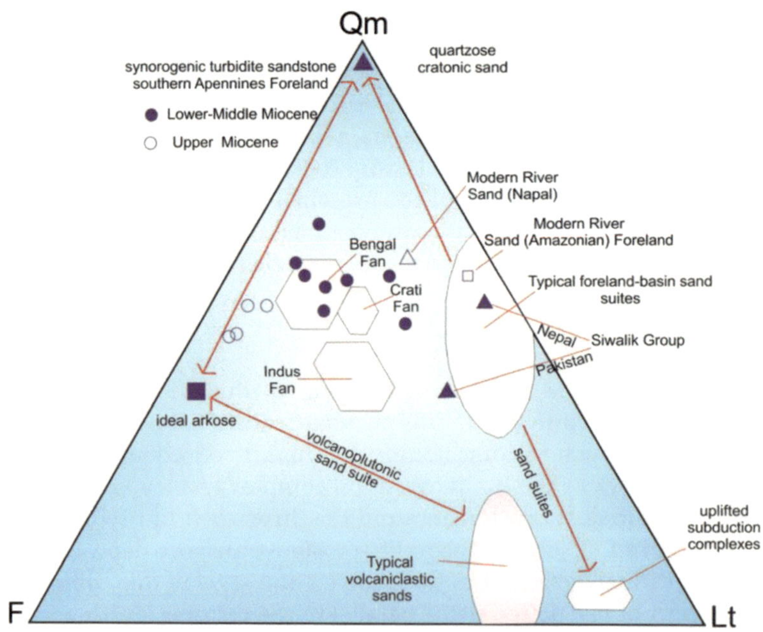

Figure 2.
QmFLt (Qm = monocrystalline quartz, F = feldspars, Lt = aphanitic lithic fragments) diagram illustrate the concept of mixing detritus from different provenance types that produce detrital modes reflecting mixed provenance [2]. Typical foreland-basin sand suites were derived from uplifted fold-thrust belts exposing sedimentary and metasedimentary strata. The mixed provenance relations are also typical of some foreland basin systems and remnant ocean basins (i.e. southern Apennines foreland, Indus and Bengal fans of the Himalayan belt). During early stage of foreland infill, sand may derive from cratonal areas, generating quartzose sand. Subsequent petrofacies is quartzolithic, and during final foreland infill (foreland uplift), petrofacies may be mixed and quartzofeldspathic. Modified after [4, 15].

foreland-basin sandstone petrofacies, testifies the complex pattern of provenance relationships during the foreland basin evolution. Quartzose sand is typical during the early stage of foreland infill, when the thrust-belt has low elevation and consequently supplies low amounts of detritus, while cratonal region is flexing and provides more amounts [33, 34]. The subsequent petrofacies is typical quartzolithic, when the thrust-belt is growing and show up its roots. Local nearby provenances from magmatic arcs, uplifted subduction complexes or uplifted carbonate rocks of the forebulge represent just small amounts of the clastic record within foreland basin system. Only if the thrust belt shows severe uplift rates, in a way to expose the crustal basement, petrofacies can evolve to quartzofeldspathic sand during the late stage of “Foreland Uplift” [3–5, 35–37].

4. Provenance and setting: Sandstone detrital modes and orogenic provenance

Foreland basins adjacent to orogenic wedges experience drastic changes in provenance during their sedimentary history due to the high stressed event. In general models, arkosic petrofacies of rift phases lie at the base of the foreland successions, followed up by quartzose petrofacies of the passive margin or cratonal regions. These pre-orogenic strata are then succeeded by quartzolithic petrofacies derived from upper crust thrusted units of the nascent orogenic system. Detrital modes in foreland setting are the combination of two main key sources located on the thrusted and uplifted upper plate, and on the underplate. These issues are crucial in the definition of key tectonostratigraphic sources that grow up from the deformational pattern of the plate convergence, closely revealed in the clastic stratigraphy of the resulting foreland basin system. Sandstone petrology is a key tool for unrevealing the close relation between source to sink and the spatial and temporal significance of the whole detrital budget.

4.1 Key tectonostratigraphic sources

4.1.1 Upper plate sources

Thin-skinned thrust terranes. - Such terranes are mainly made up of metasedimentary bedrocks and sedimentary successions directly covering their basement rock suites; other thrusted units are composed by ophiolitiferous sources of ancient consuming oceanic basins, including oceanic crust rocks of both metavolcanic and subvolcanic fragments and their oceanic sedimentary cover of chert-rich, pelagic shale and limeclasts (**Figure 3**).

Foreland Uplift terranes. - Mid-crustal deformed and thrusted terranes, including mid-high-grade metamorphic rocks (micaschist, gneiss, granulite) and plutonic suites (mainly tonalite-to-granite) are involved in the plate convergence and rapidly exhumed and uplifted. These thick-skinned thrust terranes are high producer of sand, indeed, phaneritic crustal terranes are able to generate huge volumes of feldspar-rich clastic material, and they are responsible of detrital mode shifting toward more quartzofeldspathic sandstone suites within the orogenic provenance field. Many examples along the stratigraphic record testify the changing nature of foreland sandstone petrofacies from quartzolithic to quartzofeldspathic sand suites [5, 29, 30] (**Figure 4**).

Remnant Volcanic arc. - Persistent active volcanic sources, as remnants of pre-collisional history, may contribute to the sedimentary budget of the foreland basin system when volcanic activity continues during the early stages of foreland evolution. Signatures from active volcanic sources may be diluted with other detrital

Figure 3.
Key photomicrographs of upper plate source terranes. The source terranes include: (a-to-f) obducted remnant ocean terranes of (a-b) cellular serpentinite and (c-d) serpentine-schist and eroded oceanic sedimentary cover of impure chert and argillaceous chert (e) and quartzite (f), and (g-h) folded, foliated and thrusted upper crust rock suites of low-grade metamorphic rocks of phyllite (g and h). All photos are crossed polars. Mineral abbreviations: Qm = quartz monocrystalline, Srp = serpentine schist, lm = metamorphic lithic, Lv = volcanic lithic, Phy = phyllite, Pl = plagioclase.

budget or they can represent distinctive volcaniclastic layers interbedded within the foreland clastic wedge (**Figure 5**).

4.1.2 Underplate sources

Thrusted underplate margin. - Portions of the underplate continental margin may be involved in flexure and in underthrust, and then assembled within the orogenic wedge. Usually they include sedimentary cover and upper crust stratigraphy of the underplate. Shallow-water to deep-water strata and upper crust low-grade

Figure 4.
Key photomicrographs of deeper crustal rock suites of the upper plate source terranes. The source terranes include: (a-to-c) exhumed plutonic rock suites, mainly from intermediate tonalite (a) to granodiorite (b) and granite (c), and (d-to-f) medium-high grade metamorphic rocks of garnet-bearing gneiss (d), and paragneiss (e-f). A, b, c, e, f photos are crossed polars, (e) is plane-polarized light. Mineral abbreviations: Q = quartz, Pl = plagioclase, Hnb = hornblende, K-feld = K-feldspar, Grt = garnet, lm = metamorphic lithic, CE = extrabasinal carbonate.

metamorphic rocks of the underplate continental margin generate sedimentary and metasedimentary lithic fragments within the puzzle of the foreland sandstone suites.

Forebulge. - It is the region of potential flexure uplift along the craton side of the foredeep. Because of forebulge is a positive and potentially migratory feature, which may be eroded, its potential of preservation is low. A signal of the presence of ancient forebulge may be the erosional unconformity surface. The forebulge is generally considered a zone of nondeposition or erosion, or a condensed succession, and the resulting unconformity may be used to mark its location through time. In subaerial foreland basins the forebulge is a region of erosion, with streams draining both toward and away from the orogenic belt. In submarine foreland basin systems, local carbonate patch may be developed; extensive forebulge carbonate strata can connect the foredeep with the back-bulge depozone [13]. Forebulge is a producer of sediments for the foredeep in terms of delivering detrital grains and huge gravity flow deposits. The nature of the detrital budget is in response of the underplate stratigraphic record (**Figure 6**).

Internal stable foreland. - Craton interior region of the underplate widely contributes to the initial stages of foreland generation, and usually discharge huge

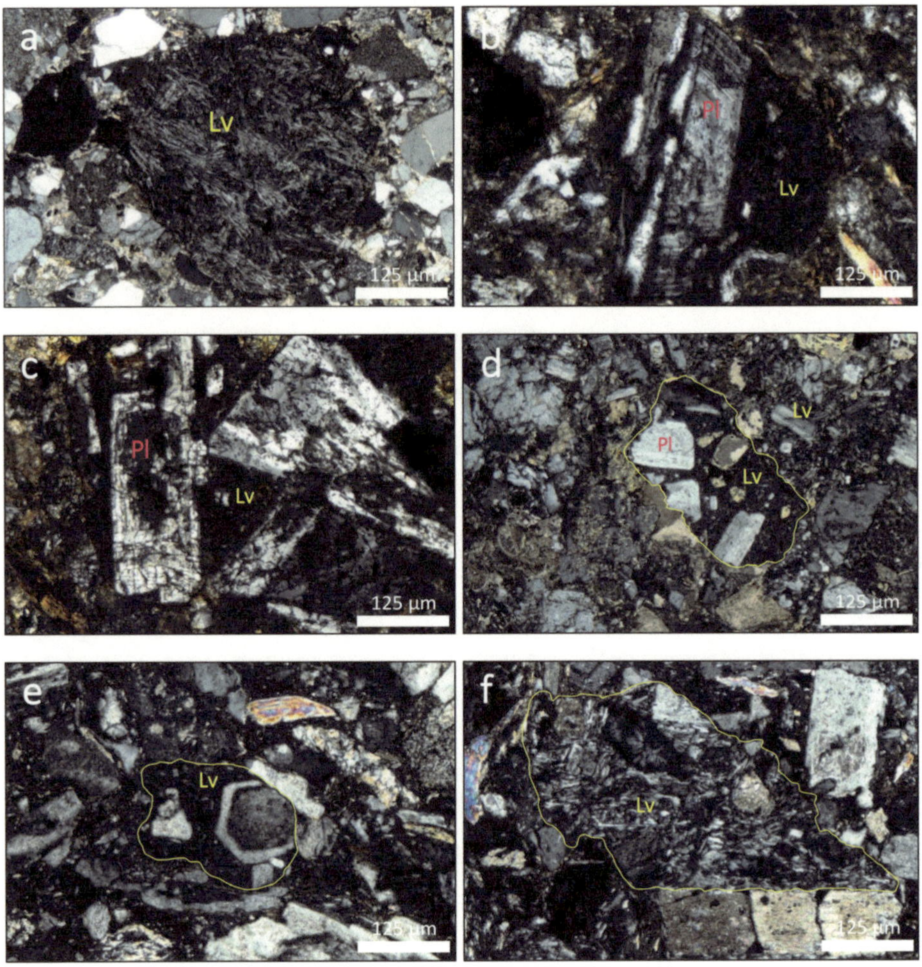

Figure 5.
Key photomicrographs of volcaniclastic signatures in foreland basins. The source terranes include: Volcanoclastic lithic rock fragment with microlithic texture (a, f) and felsitic seriate structure (b-e). All photos are crossed polars. Mineral abbreviations: Lv = volcanic lithic, Pl = plagioclase.

volumes of supermature quartzose detritus into the foreland infill. Local limeclasts and intrabasinal carbonates are an adding source (**Figure** 7).

4.2 Key sandstone petrofacies filling foreland basin

Key sandstone petrofacies reflects the changing nature of the orogenic evolution and related filling of foreland basin. Quartzose, quartzolithic, quartzofeldspathic, and hybrid sandstones are the main signatures in foreland setting (**Figure 8**). Local calclithite and volcaniclastic sandstones may also occur in this tectonic setting. Optical analysis of modal point count in sandstone is hardly suggested for the best fitting of the whole composition trying to recognize the temporal and spatial significance of the identified petrographic classes (**Table 1**).

4.2.1 Quartzose petrofacies

Quartz-rich sandstones reflect provenance from stable craton and lowlands reflecting abundance of rounded quartz grains, mainly monocrystalline. The stratigraphic record testifies occurrence of quartzose petrofacies as the main signatures

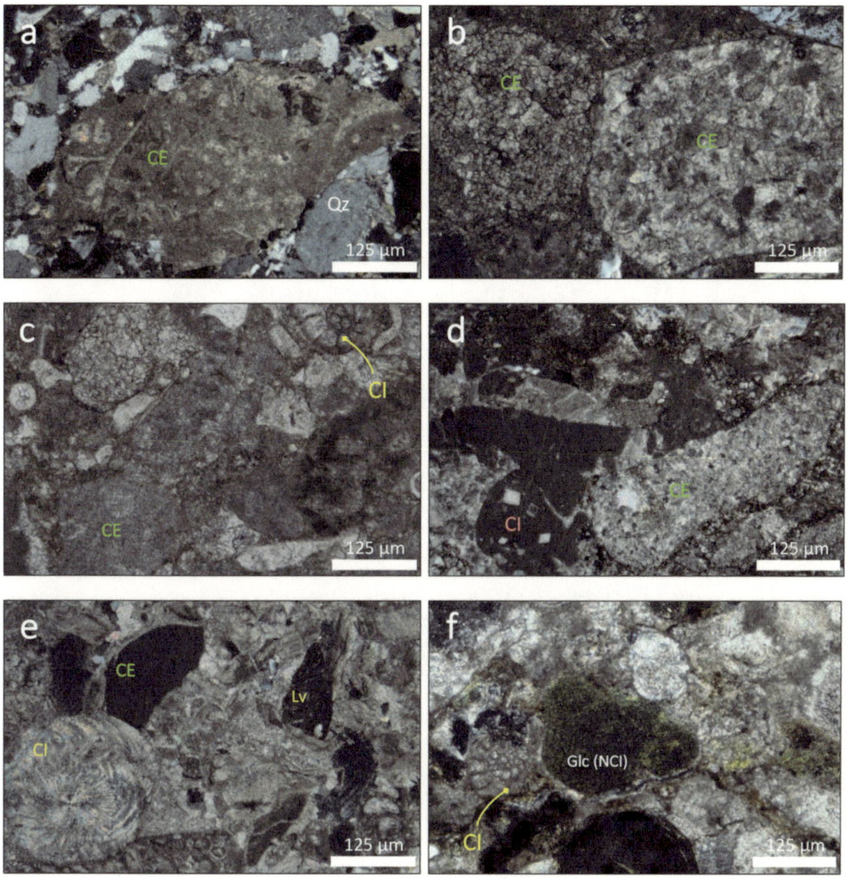

Figure 6.
Key photomicrographs of underplate source terranes. The source terranes include: (a-to-d) folded and thrusted continental margin rock suites of sedimentary cover and eventually metasedimentary bedrocks and undeformed forebulge sources. A-to-d extrabasinal carbonate grains (CE or limeclasts) in quartzolithic sandstone, calclithite and hybrid arenites. Hybrid arenites having abundant intrabasinal carbonate of bioclasts (e) and non-carbonate particles of glaucony and phosphates (f) are useful in foreland setting as related to signatures of sea-level change and underplate setting. All photos are crossed polars. A, e, f photos are crossed polars, b, c, d are in plane-polarized light. Mineral abbreviations: CE = extrabasinal carbonate, Qm = quartz, CI = intrabasinal carbonate, Lv = volcanic lithic, Glc = glaucony, noncarbonate intrabasinal (NCI).

of early stage of foreland infill [31]. Quartzose sandstone is mainly in response of downdip flexed continental margins of underplate when it represented ancient passive margins or stable craton. Great volumes of mature quartz sand can travel around the slow topography of the underplate and lay in subsiding regions of the various depozones of the foreland basin system. Circum-Mediterranean orogenic belts are a clear example of huge volumes of quartzarenite occurrence with a wide dispersal pathways all around the main orogenic fronts that accommodate more than 4,000 m of supermature quartzarenite turbidites of the Numidian Sandstone ([5] and bibliography therein]. Other examples include the Carboniferous coastal quartzarenite related to the collision of Siberia with Laurussia [36], and so many others examples in the stratigraphic record and in modern setting [40, 41].

4.2.2 Quartzolithic petrofacies

Elongated Q-L plotted quartzolithic suites is the main composition of foreland sandstones reflects the growing orogen of thin-skinned thrust belt of accreted plate. Large occurrence of metasedimentary and sedimentary lithic fragments, such as

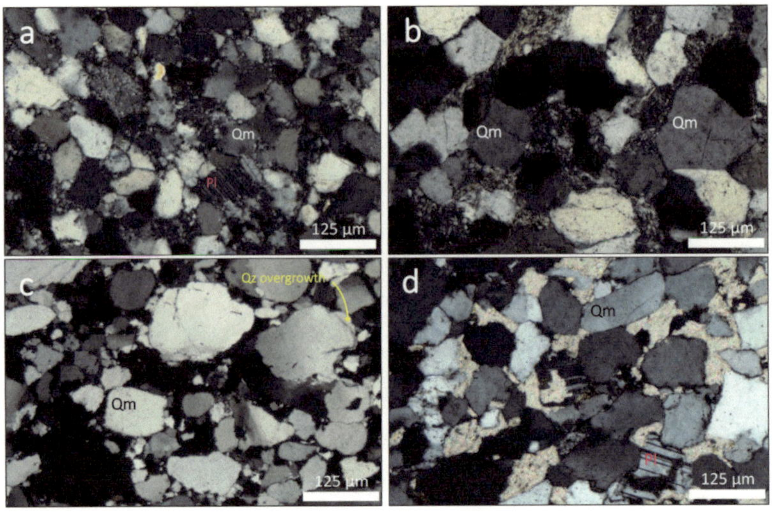

Figure 7.
Key photomicrographs of craton interior underplate provenance. The source terranes include dominant quartzose grains of internal stable foreland. (a) Quartzarenite of the antler flysch, Idaho, (b) carboniferous quartzarenite, Svalbard Island, (c) and d) well-rounded quartzarenite of the Numidian sandstone in the circum-Mediterranean orogenic region. All photos are crossed polars.

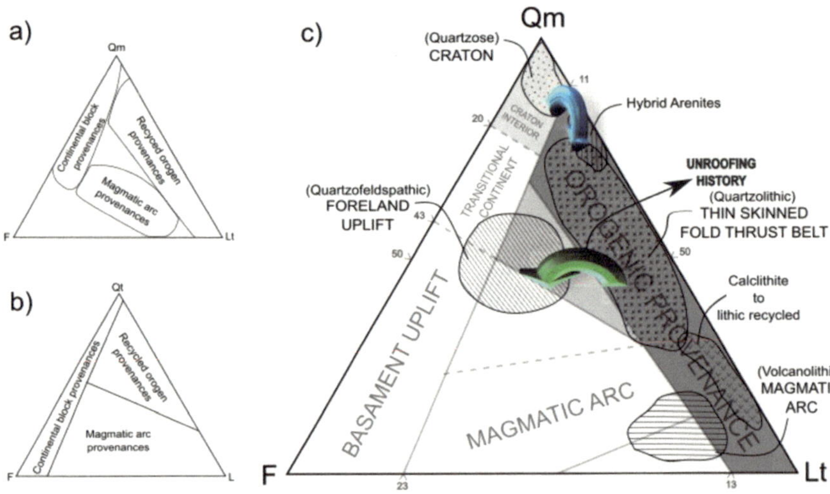

Figure 8.
QmFLt (Qm = monocrystalline quartz, F = feldspars, Lt = aphanitic lithic fragments) diagram used for illustrate the concept of different provenance types in relation to tectonic setting. (a) Sources for original conception of plate tectonic setting and sandstone composition by Dickinson and Suczek [38], and (b) later refinement by Dickinson [1]. (c) Typical foreland-basin sand suites were derived during early to late stages from: (i) craton interior, (ii) upper crust thrusted (iii) interbedded volcaniclastic suites from remnant magmatic arc, and (iv) uplifted mid-crustal fold-thrust belts. During early stage of foreland infill, sand may derive from cratonal areas, generating quartzose sand. Subsequent petrofacies is quartzolithic, and during final foreland infill (foreland uplift), petrofacies may be mixed and quartzofeldspathic testifying the unroofing history of the Orogenic Belt and related foreland infill (arrow).

abundant quartz, are the main detrital supply to the foreland basin. Nice examples are quartzolithic sandstones of the Eocene-to-Pliocene foreland basins of the Himalayan thrust-belt [3, 35], as such as many other sandstone examples of Paleozoic through Pleistocene orogens [5, 11, 15]. On both ancient and recent foreland sand (stone), the main detrital component is the abundance of quartz and aphanitic lithic fragments. The latter are mainly represented by (i) low-to-medium grade

Petrographic classes					
		Quartz (Qt=Qm+Qp)			**Extrabasinal Carbonates (CE)**
NCE	Qm	Quartz (single crystals)	CE	Ls	Dolostone
	Qp	Polycrystalline quartz with tectonic fabric			Micritic Limestone
		Polycrystalline quartz without tectonic fabric			Sparitic Limestone
	Qm	Quartz in metamorphic r.f.			Microsparitic Limestone
		Quartz in volcanicnic r.f.			Biomicritic Limestone
		Quartz in plutonic or gneissic r.f.			Biosparitic Limestone
		Calcite replacement on quartz			Fossil (single skeleton)
					Fossil in Limestone-Dolostone
		Feldspars (F=K+P)			Single spar (calcite)
	K	K-feldspar (single crystals)			Single spar (dolomite)
		K-feldspar in plutonic r.f.			Fe-Oxid concretions
		Calcite replacement in k-feldspar			**Intrabasinal Carbonates (CI) and noncarbonates (NCI)**
	P	Plagioclase (single crystals)		CI	Bioclast
		Plagioclase in metamorphic r.f.			Peloids
		Plagioclase in plutonic r.f.		NCI	Glauconite
		Plagioclase in plutonic or gneissic r.f.			Oxide-Fe
		Plagioclase in sandstone			Rip-Up clasts
		Plagioclasio in volcanic r.f.			
		Calcite replacement in Plagioclase			**Interstitial components (Matrix and cements)**
					Siliciclastic matrix
		Micas			Carbonate matrix (micrite)
		Micas and clorite (single crystals)			Carbonate cement (pore-filling)
		Micas in volcanic r.f.			Carbonate cement (patchy calcite)
		Micas in plutonic r.f.			Calcite replacement on underterm. grain
		Micas in metamorphic r.f.			Siliceous cement
					Albite cement
		Lithic fragments (L=Lm+Lv +Ls)			Phyllosilicate cement
	Lv	Volcanic lithic with microlithic fabric			Oxid-Fe cement
		Volcanic lithic with felsitic granular texture			Quartz overgrowth
		Volcanic lithic with felsitic seriate texture			Clay and Clay grain coats
		Volcanic lithic with lathwork texture			Alterites (indeterm.alterite grain)
		Volcanic lithic with vitric texture			

	Petrographic classes	
	Quartz (Qt=Qm+Qp)	**Extrabasinal Carbonates (CE)**
	Serpentinite	
	Serpentine-schist	
	Phyllite	
Lm	Fine-grained Schist	
	Slate	
	Siltstone	
	Impure Chert	
Ls	Shale	
	Dense minerals	
	Dense mineral (single crystal)	
	Dense mineral in plutonic r.f.	
	Dense mineral in volcanic r.f.	
	Dense mineral in metamorphic r.f.	
	Opaque minerals	

Table 1.
Optical petrographic classes for modal analysis of arenites by using temporal and spatial criterion of the various modal classes [39–41]. NCE: noncarbonate extrabasinal grains; CE: carbonate extrabasinal grains; NCI: noncarbonate intrabasinal grains; CI: carbonate intrabasinal grains.

metamorphic lithic fragments of phyllite, quartzite, micaschist, and metavolcanics; (ii) ophiolitiferous detritus of obducted oceanic crust, including serpentinite and serpentine schist, lathwork textures of volcanic fragments of basalts, sedimentary chert, argillaceous chert and pelagic limestone; (iii) sedimentary lithic fragments of sedimentary strata involved in the orogenic wedge, including siliciclastics (siltstone, shale), chert, and extrabasinal carbonate grains; (iv) volcanic lithic fragments of both older volcanic suites and coeval volcanic sources (see section 4.2.4).

4.2.3 Quartzofeldspathic petrofacies

Plotted in the intermediate portions of the QFL diagram, sandstones having abundance of quartz and feldspar and minor lithic fragments are quartzofeldspathic sandstone suites. They reflect the uplift foreland stage of late orogenic phases when mid-crustal rocks (dominantly high-grade metamorphic and plutonic suites) are involved in thrusting, exhumation and uplift. Key lithostratigraphic units in the geological record have quartzofeldspathic sandstone suites as Old Red Sandstone of the Caledonian orogen [36] or New Red Sandstone of the Variscan orogen (e.g. Val Gardena Fm.) [42], as such as late petrofacies units of many other orogens [5, 35].

4.2.4 Interbedded volcaniclastic petrofacies

Coeval volcaniclastic sandstones may accommodate in foreland region due to persistent volcanism, related to the subduction during closure of remnant ocean basins and early foreland stage. Dominantly arc-derived volcaniclastic sandstones occur in foreland setting as important detrital contribution in quartzolithic suites and as interbedded volcaniclastic suites, related to the coeval volcanism with sedimentation. Volcanic debris is represented mainly as abundance of vitric fragments (ash shards) in ash turbidites or as occurrence of typical textural attributes in volcanic lithic fragments [3]. The most typical volcanic textures in foreland strata are vitric, felsitic granular, felsitic seriate and microlithic reflecting typical silicic volcanism from trachite-andesite to rhyodacite and rhyolite.

4.2.5 Hybrid arenite petrofacies

Hybrid arenites are defined by Zuffa [39, 43–45] as the mixing of extrabasinal detrital grains, both siliciclastics and carbonate, and intrabasinal detrital grains, both noncarbonate and carbonate. In many submarine foreland settings, the mixing of intrabasinal and extrabasinal detrital grains reflects hybrid arenite suites. Mostly is the combination of siliciclastic particles together with the main allochemical grains [46] or intrabasinal carbonate grains. These signatures are in response of remobilized intrabasinal grains during huge arrival of subaerial clastics, temporally stored in shallow water. Deep-water turbidite systems in foreland basins include large occurrence of hybrid arenites suggesting the active rules of shallow-marine environment. Also, non-carbonate -rich (mainly glaucony and phosphatic) hybrid arenite signatures testify maximum flooding during sea-level changes [47].

4.2.6 Calclithite petrofacies

Limeclasts [38] and extrabasinal carbonate grains [39, 43, 44] signatures are typical in foreland settings [48, 49]. Ancient carbonate grains derive from eroded older carbonate strata and reflect, in foreland setting, diverse but interfingered dispersal pathways drained from the fold-thrust belt, intrabasinal structural highs, forebulge and the stable foreland. Distinguishing the nature and location of older carbonate detritus is crucial for detailed paleogeographic reconstructions. Older carbonate source rocks may occur as both (i) subaerial exposed carbonate strata interbedded within the fold-thrust belt, and as (ii) intrabasinal structural highs within wedge-top and foredeep depozones, that can result as base-of-slope carbonate breccia and related turbidite sandstones. Within the circum-Mediterranean region, ancient foreland sandstone strata reflect important contributions from older carbonate detritus in the Pyrenees [50], Betic Cordillera [40, 51], Iberian Range [52, 53], Alps and Apennines [5, 48, 49, 54], due to large abundance of Mesozoic-to-Cenozoic carbonate platform/to basinal successions related to the Neo-Tethyan oceanic rifting.

4.3 Spatial and temporal significance of sand grains

More refined compositional signatures of sand (stone) in foreland setting include temporal (coeval vs. noncoeval) and spatial (extrabasinal vs. intrabasinal) decoding of clastic particles [39, 44]. The large spectrum of detrital grains in sand (stone) has high value in inferring the spatial/temporal constraints. Carbonate and volcanic detrital grains are particularly sensitive of these discriminant subdivisions if the ultimate goal is the correct palaeogeographic reconstructions. For instance, carbonate grains can be spatially generated in extrabasinal and intrabasinal environments, and can be noncoeval (paleo) or coeval (neo) with sedimentation: (i) noncoeval extrabasinal carbonate grains are eroded and grains generate carbonate lithic fragments (dolostone and limestone); (ii) coeval extrabasinal carbonate grains are fragments of nowely formed carbonate concrections in soils (calcrete, caliche) or travertine fragments; (iii) coeval intrabasinal carbonate grains are the typical allochemical particles [46] including ooids, bioclasts, intraclasts and peloid; (iv) noncoeval intrabasinal carbonate grains include older carbonate strata in structural-related intrabasinal highs that can deliver large blocks and sand at the base of slope.

Volcanic particles represent the most intricate task in optical analysis for the discrimination between grains eroded from ancient volcanic rocks (paleovolcanic or noncoeval grains) and grains generated by intrabasinal or extrabasinal active volcanism during sedimentation (neovolcanic, coeval grains) [39, 44, 55]. Apart the contributions from older eroded volcanic rocks that refine the general puzzle of source areas, the coeval volcaniclastic contribution represent a well-defined marker in the sedimentary record. Volcanic particles can reveal important constraints on deciphering spatial (extrabasinal vs. intrabasinal) and temporal relationships of neovolcanic events (pre-, syn-, inter- and post-eruptive periods).

5. Summary

The whole chapter tries to summarize the typical fold thrust belt and foreland basin system settings from the petrological point of view of the key sandstone suites. This kind of reasoning permits to point out all the evolutionary events of these tectonic frameworks. After explaining the teorethical key concepts of the tectonic structures involved and the main provenance models, the discussion points rawly to the key sandstone petrofacies that commonly represent the infilling of the basin settings. The early stage of the foreland usually reflects quatzose sandstone suites derived from cratonal area. The huge volumes of the foreland infill has quartzolithic sandstone suits, derived from the growing orogenic belt. Finally quatzofeldspathic sandstone compositions are mainly derived from uplifted thrust belt. Local volcaniclastic and hybrid arenites can also occur during the foreland history.

6. Conclusions

In foreland settings, subsidence and uplift are profoundly affected by lithospheric flexure. Foreland basin subsidence is primarily controlled by downflexing of the lithosphere in response to thrust accommodation and loading. The interrelationships between lithospheric flexure, single thrust accommodation within the accretionary wedge and flexural subsidence experiences geometrically complexes entities within the foreland region [4, 5]. Sandstone signatures of foreland-related basins reflect the close changing nature of detrital composition during the growth of the orogenic belt and the flexure of the underplate. Quartzolithic sand (stone) suites are the main composition of foreland sandstones during syntectonic evolution of foreland infill. However, quartzose suites occur during the early stage of foreland infill, as such as quartzofeldspathic sandstones during the later foreland uplift. Local contributions of volcanolithic sandstones from magmatic arc may occur as remnant arc activity continues during initial orogenic processes. Finally, carbonate-rich strata derived from both erosion of ancient carbonate successions and coeval carbonate/noncarbonate detritus are typical in many marine foreland infill.

In conclusions, close relations between regional sandstone petrofacies and major unconformities may contribute to the more precise paleogeographic and paleotectonic reconstructions.

Acknowledgements

This chapter was supported by the "Laboratory of Geodynamics, Geo-Paleobiology and Earth Surface Processes" of the Department of Biology, Ecology and Earth Sciences at the University of Calabria. This research was funded by MIUR (Italian Ministry of Education, University and Research), MIUR-PRIN Project 2006.04.8397 "The Cenozoic clastic sedimentation within the circum-Mediterranean orogenic belts: implications for palaeogeographic and palaeotectonic evolution", the 2010-2012 MIUR-PRIN Project The Thrace sedimentary basin (Eocene-Quaternary; Turkey, Greece, Bulgaria): stratigraphic depositional architecture and sediment dispersal pathway within postorogenic basins", and the University of Calabria MIUR-ex60% Projects (Palaeogeographic and Palaeotectonic Evolution of the Circum-Mediterranean Orogenic Belts, 2001–2005; and Relationships between Tectonic Accretion, Volcanism and Clastic Sedimentation within the Circum-Mediterranean Orogenic Belts, 2006-2018; support to S. Critelli). Sara Criniti was supported by The PhD Course in "Science and Engineering of Environment, Building and Energy" at the University of Calabria.

Author details

Salvatore Critelli* and Sara Criniti
Department of Biology, Ecology and Earth Sciences, University of Calabria, Rende, CS, Italy

*Address all correspondence to: salvatore.critelli@unical.it

References

[1] Dickinson W.R., 1985. Interpreting provenance relations from detrital modes of sandstones. In: G.G. Zuffa (ed.), Provenance of Arenites, 333–361. D. Reidel, Dordrecht.

[2] Dickinson, W.R., 1988. Provenance and sediment dispersal in relation to paleotectonics and paleogeography of sedimentary basins. In: K.L. Kleinspehn, C. Paola (eds.), New Perspectives in Basin Analysis, 3–25. New York, Springer-Verlag.

[3] Critelli S., Ingersoll R.V., 1994. Sandstone petrology and provenance of the Siwalik Group (northwestern Pakistan and western-southeastern Nepal). J. Sedim. Res., A64: 815–823.

[4] Critelli, S., 1999. The interplay of lithospheric flexure and thrust accommodation in forming stratigraphic sequences in the southern Apennines foreland basin system, Italy. Accademia Nazionale dei Lincei, Rendiconti Lincei Scienze Fisiche e Naturali 10, 257–326.

[5] Critelli, S., 2018. Provenance of Mesozoic to Cenozoic Circum-Mediterranean sandstones in relation to tectonic setting. Earth-Science Reviews, v. 185, p. 624–648.

[6] Beaumont C., 1981. Foreland Basins. Geoph. J. Royal Astr. Soc., 65: 291–329.

[7] Sinclair H.D., Allen P.A., 1992. Vertical versus horizontal motions in the Alpine orogenic wedge: stratigraphic response in the foreland basin. Basin Res., 4: 215–232.

[8] Dickinson W.R., 1974. Plate tectonics and sedimentation. In: W.R. Dickinson (ed.), Tectonics and Sedimentation. SEPM (Soc. for Sedim. Geol.), spec. pub. 22: 1–27.

[9] Bally A.W., Snelson S., 1980. Realms of subsidence. Mem. Can. Soc. Petrol. Geol., 6: 9–75.

[10] Allen, P.A., Homewood, P.H., and Williams, G.D., 1986. Foreland basins: an introduction. In: P.A. Allen, P. Homewood (eds.), Foreland Basins. Int. Ass. Sediment., spec. pub. 8, p. 3-12.

[11] Miall A.D., 1995. Collision-related foreland basins. In: C.J. Busby, R.V. Ingersoll (eds.), Tectonics of Sedimentary Basins, 393–424. Blackwell Science, Oxford.

[12] DeCelles P.G., Giles K.A., 1996. Foreland Basin Systems. Basin Research, 8: 105–123.

[13] Giles K.A., Dickinson, W.R., 1995. The interplay of eustasy and lithospheric flexure in forming stratigraphic sequences in foreland settings: an example from the Antler foreland, Nevada and Utah. In: S.L. Dorobek, G.M. Ross (eds.), Stratigraphic Evolution of Foreland Basins. SEPM (Soc. for Sedim. Geology) spec. pub. 52, 187–211.

[14] Critelli, S., Muto, F., Tripodi, V., Perri, F., 2011. Relationships between lithospheric flexure, thrust tectonics and stratigraphic sequences in foreland setting: the Southern Apennines foreland basin system, Italy, *in* Schattner, U., ed., New Frontiers in Tectonic Research at the Midst of Plate Convergence. Intech Open Access Publisher, Janeza Trdine 9, Rijeka, Croatia, 121–170.

[15] Critelli, S., Muto, F., Perri, F., Tripodi, V., 2017. Interpreting Provenance Relations from Sandstone Detrital Modes, Southern Italy Foreland Region: stratigraphic record of the Miocene tectonic evolution. Marine and Petroleum Geology, v. 87, p. 47–59.

[16] Jordan T.E., 1981. Thrust loads and foreland basin evolution, Cretaceous, western United States. Am. Ass. Petrol. Geol. Bull., 65: 291–329.

[17] Schedl A., Wiltschko D.V., 1984. Sedimentological effects of a moving terrain. J. Geology, 92: 273–287.

[18] Peper T., Van Balen R., Cloetingh S., 1995. Implications of orogenic wedge growth, intraplate stress variations, and eustatic sea-level change for foreland basin stratigraphy - inferences from numerical modeling. In: S.L. Dorobek, G.M. Ross (eds.), Stratigraphic Evolution of Foreland Basins. SEPM (Soc. for Sedim. Geology) spec. pub. 52, 25–35.

[19] Jordan T.E., 1995. Retroarc foreland and related basins. In: C.J. Busby, R.V. Ingersoll (eds.), Tectonics of Sedimentary Basins. Blackwell Science, pp. 331–362.

[20] Quinlan G.M., Beaumont C., 1984. Appalachian thrusting, lithospheric flexure and Paleozoic stratigraphy of the eastern interior of North America. Can. J. Earth Sc., 21: 973–996.

[21] Vai, G.B., 1987. Migrazione complessa del sistema fronte deformativo-avanfossa-cercine periferico: il caso dell'Appennino settentrionale. Mem. Soc. Geol. It., 38: 95–105.

[22] Wiltschko D.V., Door J.A. Jr., 1983. Timing of deformation in overthrust belt and foreland of Idaho, Wyoming, and Utah. American Association of Petroleum Geologists Bulletin, v. 67, p. 1304–1322.

[23] Graham S.A. et 14 others, 1986. Provenance modelling as a technique for analysing source terrane evolution and controls on foreland sedimentation. In: P.A. Allen, P. Homewood (eds.), Foreland Basins. Int. Ass. Sediment., spec. pub. 8, p. 425–436.

[24] Jordan T.E., Flemings P.B., Beer J. A., 1988. Dating thrust-fault activity by use of foreland-basin strata. In: K.L. Kleinsphen, C. Paola (eds.), New Perspectives in Basin Analysis, p. 307–330. Springer-Verlag, New York.

[25] Steidtmann J.R. and Schmitt J.G., 1988. Provenance and dispersal of tectogenic sediments in thin-skinned, thrusted terrains. In: K.L. Kleinsphen, C. Paola (eds.), New Perspectives in Basin Analysis, 353–366. Springer-Verlag, New York.

[26] DeCelles P.G., 1988. lithologic provenance modeling applied to the Late Cretaceous Echo Canyon Conglomerate, Utah: a case of multiple source areas. Geology, v. 19, p. 1039–1043.

[27] Colombo F., 1994. Normal and reverse unroofing sequences in syntectonic conglomerates as evidence of progressive basinward deformation. Geology, 22: 235–238.

[28] Critelli S., Le Pera E., 1994. Detrital modes and provenance of Miocene sandstones and modern sands of the southern Apennines thrust-top basins (Italy). J. Sedim. Res., A64: 824–835.

[29] Critelli S., Le Pera E., 1995. Tectonic evolution of the Southern Apennines thrust-belt (Italy) as reflected in modal compositions of cenozoic sandstone. J. Geology, 103: 95–105.

[30] Critelli S., Le Pera E., 1998. Post-Oligocene sediment-dispersal systems and unroofing history of the Calabrian microplate, Italy. Int. Geol. Rev., 40: 609–637.

[31] Schwab F.L., 1986. Sedimentary "signatures" of foreland basin assemblages: real or counterfeit?. In: P. A. Allen, P. Homewood (eds.), Foreland Basins. Int. Ass. Sediment. spec. pub. 8: 395–410.

[32] DeCelles P.G., Hertel F., 1989. Petrology of fluvial sands from the Amazonian foreland basin, Peru and Bolivia. Geol. Soc. Am. Bull., 101: 1552–1562.

[33] Dickinson, W.R., Lawton, T.F., and Inman, K.F., 1986. Sandstone detrital modes, central Utah foreland region: stratigraphic record of cretaceous-Paleogene tectonic evolution. Journal of Sedimentary Petrology, v. 56, p. 276–293.

[34] Cazzola, C., Critelli, S., 1987. Litostratigrafia e petrologia delle quarzoareniti torbiditiche oligomioceniche di Asilah (Catena del Rif, Marocco nord-occidentale). Mineralogica et Petrographica Acta, 30, 203–226.

[35] Garzanti E., Critelli S., Ingersoll R. V., 1996. Paleogeographic and paleotectonic evolution of the Himalayan Range as reflected by detrital modes of Tertiary sandstones and modern sands (Indus Transect, India and Pakistan). Geol. Soc. Am. Bull., 108: 631–642.

[36] Critelli S., Reed W.E., 1999. Provenance and stratigraphy of the Devonian (Old Red Sandstone) and Carboniferous sandstones of Spitsbergen, Svalbard. European J. Mineralogy, 11: 149–166.

[37] Barone M., Dominici R., Muto F., Critelli S., 2008. Detrital modes in a late Miocene wedge-top basin, northeastern Calabria, Italy: Compositional record of wedge-top partitioning. Journal of Sedimentary Research, 2008, v. 78, 693–711.

[38] Dickinson, W.R., Suczek, C.A., 1979. Plate tectonics and sandstone compositions. American Association of Petroleum Geologists Bulletin, v. 63, p. 2164–2183.

[39] Zuffa G.G., 1987. Unravelling hinterland and offshore paleogeography from deep-water arenites. In: J.K. Leggett, G.G. Zuffa (eds.), Deep-Marine Clastic Sedimentology. Concepts and Case Studies, 39–61. London, Graham and Trotman.

[40] Critelli, S., Arribas, J., Le Pera, E., Tortosa, A., Marsaglia, K.M., Latter, K. K., 2003. The recycled orogenic sand provenance from an uplifted thrust-belt, Betic Cordillera, southern Spain. Journal of Sedimentary Research, v. 73, 72–81.

[41] Critelli S., Le Pera E., 2003. Provenance relations and modern sand petrofacies in an uplifted thrust-belt, northern Calabria, Italy. In: Quantitative Provenance Studies in Italy (Ed. by R. Valloni & A. Basu). Servizio Geologico Nazionale, Memorie Descrittive della Carta Geologica d'Italia, v. 61, p. 25–39.

[42] Fontana, D., and Zuffa, G.G., 1982. Composizione e provenienza delle Arenarie di Val Gardena e del Verrucano lombardo (Permiano, Alpi). Memorie della Società Geologica Italiana, v. 24, p. 51–64.

[43] Zuffa, G.G., 1980. Hybrid arenites: Their composition and classification. Journal of Sedimentary Petrology 50, 21–29.

[44] Zuffa, G.G., 1985. Optical analysis of arenites: influence of methodology on compositional results, in Zuffa, G.G., ed., Provenance of Arenites, 165–189. D. Reidel, Dordrecht.

[45] Zuffa, G.G., 1991, On the use of turbidite arenites in provenance studies: Critical remarks, in Morton, A.C., Todd, S.P., and Haughton, P.D.W., eds., Developments in Sedimentary Provenance Studies: Geological Society [London] Special Publication 57, p. 23–29.

[46] Folk, R.L., 1974. Petrology of Sedimentary Rocks: Austin, Texas, Hemphill Publishing Company, 182 p.

[47] Amorosi, A., 1995. Glaucony and sequence stratigraphy: A conceptual framework of distribution in siliciclastic sequences: Journal of Sedimentary Research, v. B65, p. 419–425.

[48] Fontana, D., 1991. Detrital carbonate grains as provenance indicators in the Upper Cretaceous Pietraforte Formation (northern Apennines). Sedimentology, 38, 1085–1095.

[49] Critelli, S., Le Pera, E., Galluzzo, F., Milli, S., Moscatelli, M., Perrotta, S., Santantonio, M., 2007. Interpreting siliciclastic-carbonate detrital modes in foreland basin systems: an example from upper miocene arenites of the central Apennines, Italy. In: Arribas, J., Critelli, S., Johnsson, M. (Eds.), Sedimentary Provenance: Petrographic and Geochemical Perspectives. Geological Society of America Special Paper 420, pp. 107–133.

[50] Fontana, D., Zuffa, G.G., and Garzanti, E., 1989, The interaction of eustasy and tectonism from provenance studies of the Eocene Hecho Group turbidite complex (south-central Pyrenees, Spain): Basin Research, v. 2, 223–237.

[51] Critelli, S., Martin-Martin, M., Capobianco, W., Perri, F., 2021. Sedimentary history and palaeogeography of the Cenozoic clastic wedges of the Malaguide Complex, Internal Betic Cordillera, southern SpainMarine and Petroleum Geology, v. 124, p. 1–21 (104775).

[52] Arribas, J., Critelli, S., Le Pera, E. and Tortosa, A., 2000. Composition of modern stream sand derived from a mixture of sedimentary and metamorphic source rocks (Henares River, Central Spain). Sedimentary Geology, v. 133, 27–48.

[53] Arribas, M.E., and Arribas, J., 2007. Interpreting carbonate particles in modern continental sands: An example from fluvial sands (Iberian range, Spain). In: Arribas, J., Critelli, S., Johnsson, M. (Eds.), Sedimentary Provenance: Petrographic and Geochemical Perspectives. Geological Society of America Special Paper 420, pp. 167–179.

[54] Cibin, U., Spadafora, E., Zuffa, G. G., Castellarin, A., 2001. Continental collision history from arenites of episutural basins in the northern Apennines, Italy. Geological Society of America Bulletin, 113, 4–19.

[55] Critelli S., Ingersoll R.V., 1995. Interpretation of neovolcanic versus palaeovolcanic sand grains: an example from Miocene deep-marine sandstone of the Topanga Group (southern California). Sedimentology, 42: 783–804.

Chapter 2

Stylolite in Upper Cretaceous Carbonate Reservoirs from Northwestern Iraq

Ali Al-Juboury, Mohammed A. Al-Haj and Aboosh H. Al-Hadidy

Abstract

Stylolites are commonly observed in the carbonate reservoirs in various oilfield of Iraq including those of upper Cretaceous successions from northwestern Iraq, where they are characterized by stylolite-rich zones in the Cenomanian-early Turonian Gir Bir Formation and to a lesser extent in the Turonian-Santonian Wajna and early Campanian Mushorah formations respectively. The observed stylolites are either large to be identified in the core samples or smaller ones that are well observed in the thin sections and are characterized by variations in amplitude, morphology and accumulated insoluble residues. The recorded stylolites are classified as hummocky, irregular, low and high-amplitudes peaks, and irregular anastomosing stylolites. Stylolites affect the porosity permeability and thickness reduction compaction as the main chemical compaction (pressure solution) that reduce porosity. Whereas, in other places, the stylolites act as seals and stop the upward movement of hydrocarbons. This is also seen for mineralization processes such as silicification that ended near the stylolite surfaces.

Keywords: Stylolites, Solution seams, Hydrocarbon movement, Carbonate reservoir, Cretaceous, Iraq

1. Introduction

The largest hydrocarbon reserves in Iraq are hosted in the Cretaceous, particularly in the Mesopotamian Basin which are composed mainly of carbonate rocks [1]. The Cretaceous succession in Iraq reaches up to 3000 m thick and represents part of the megasequences AP8 and AP9 of the Arabian plate sequence stratigraphy which also forms one of richest hydrocarbon provinces of the world [2, 3]. These carbonate successions include various pressure solution features including stylolites and solution seams. Stylolite is a pressure solution feature commonly observed in carbonate reservoir rocks and usually results in reduction of unit thickness and porosity and permeability of the reservoirs [4–6]. Stylolites commonly form along lithologic transitions or partings as a result of stress due to either overburden pressure or tectonism [6]. In the present work, stylolites in the lower to upper Cretaceous carbonates in the reservoirs from northwestern Iraq have been studied in Gir Bir (Cenomanian-early Turonian), Wajna (Turonian-Santonian) and Mushorah (early Campanian) formations (**Figure 1**) in order to discuss the main types of stylolites and their effect on reservoir quality of the studied carbonates.

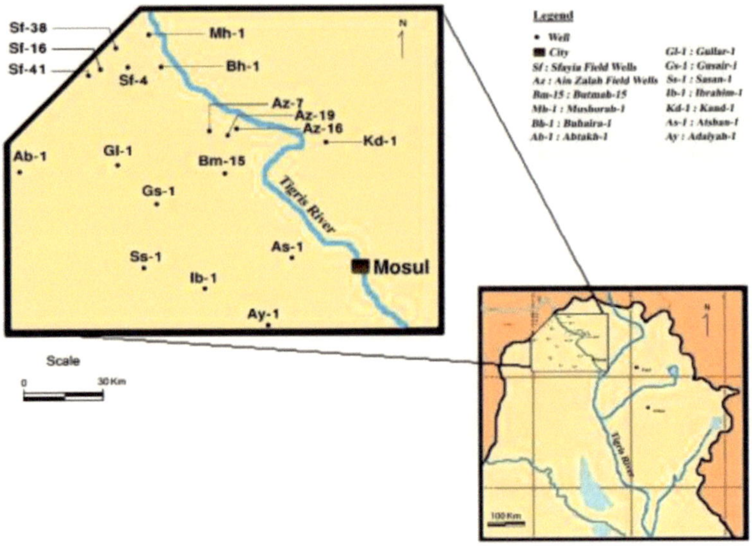

Figure 1.
Location of the studied wells, northwestern Iraq.

2. Geologic setting

Carbonates rocks that form the main lithology in the Cretaceous successions in Iraq and most of the Middle East region are formed mainly due to prevalence of warm equatorial climates [7] leading to deposition of reefal and lagoonal facies of the Gir Bir Formation, the outer lagoonal (shoal) facies of the Wajna Formation and the outer shelf to upper bathyal facies of the Mushorah Formation in the area of study of northwestern Iraq [8]. Tectonically the area of study lies in the Chamchamal-Butmah subzone of the Foot Hill zone of Iraq [9] which is a part of the Low Folded Zone of Iraq (see **Figure 2**) that have been eroded in the upper Cretaceous as a result of several phases of tectonic movements [11]. Therefore, the area was separated due to the effect of Mosul and Khleisia uplift from paleoneotethys main basin.

According to Sharland et al. [3], the studied formations lie in two of Arabian Plate Megasequences namely AP8 and AP9. The Gir Bir Formation represents the upper part of the late Tithonian-early Turonian megasequence AP8, whereas, the Wajna and Mushorah formations lie in the lower part of late Turonian-Danian megasequence AP9. Due to uplifting in the region, exposure of the shelf led to regional unconformity surface which represents a huge hiatus covering most part of northern Iraq [12] and extends to other parts of the Arabian Plate at the top of AP8 [3], (**Figure 3**). Two reasons were suggested to this unconformity surface; local uplifting of the region followed by obduction and emplacement of the ophiolite in northeastern parts of Iraq [2, 7] or global eustatic decline of sea level in the end of late Turonian [14, 15].

3. Materials and methods

Petrographic study on the carbonate succession from the three studied formations are conducted in order to determine their main petrographic constituents and the

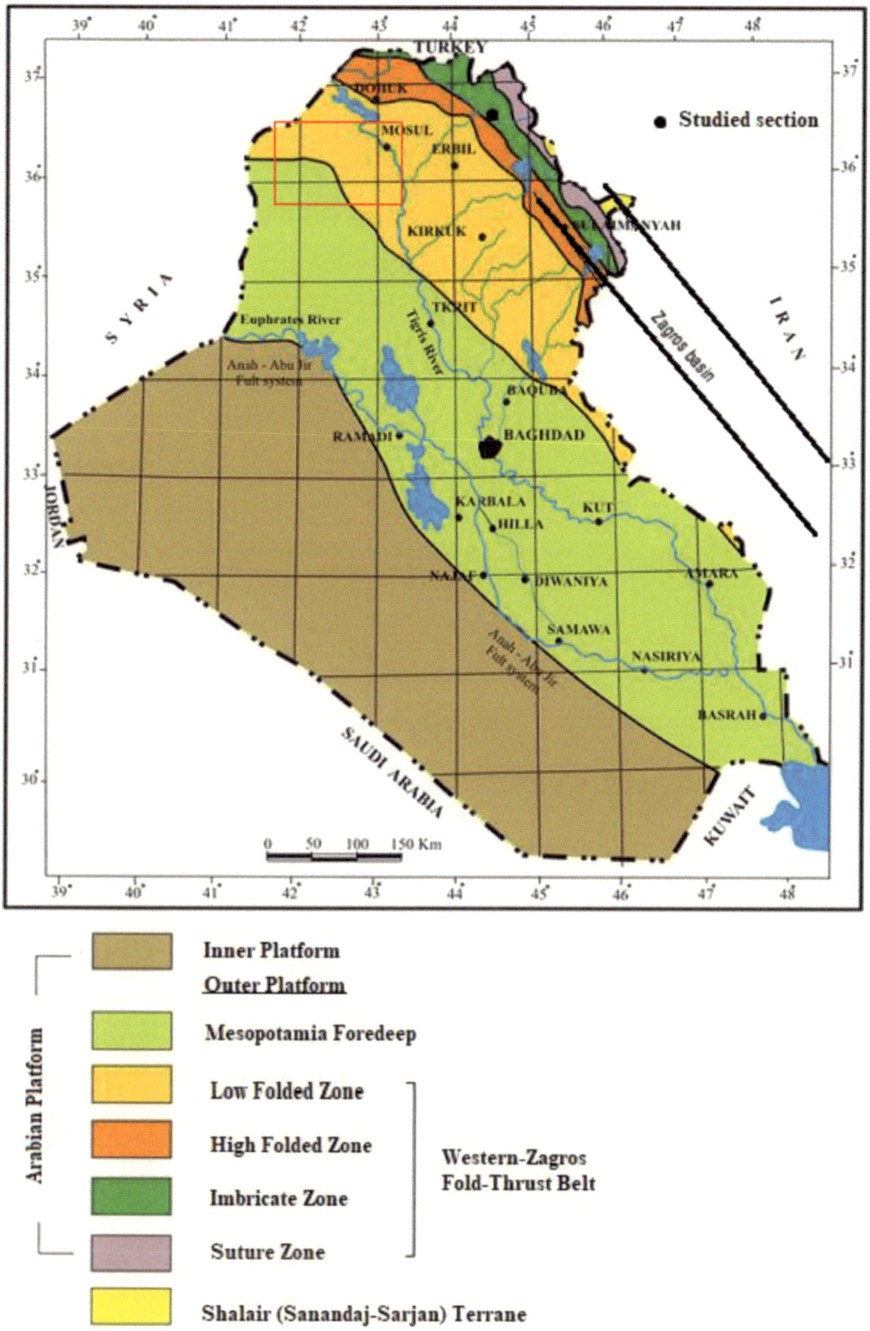

Figure 2.
Tectonic divisions of Iraq and the location of the study area which is marked with a dashed red rectangle [10].

diagenetic processes affecting on the studied carbonates including the fractures and stylolites. This study is performed at the Geology Department of Mosul University, Iraq. SEM analysis on selected samples was conducted at the Steinmann Institute of Bonn University, Germany using a Camscan MV 2300 SEM with a calibrated energy dispersive X-ray analysis system.

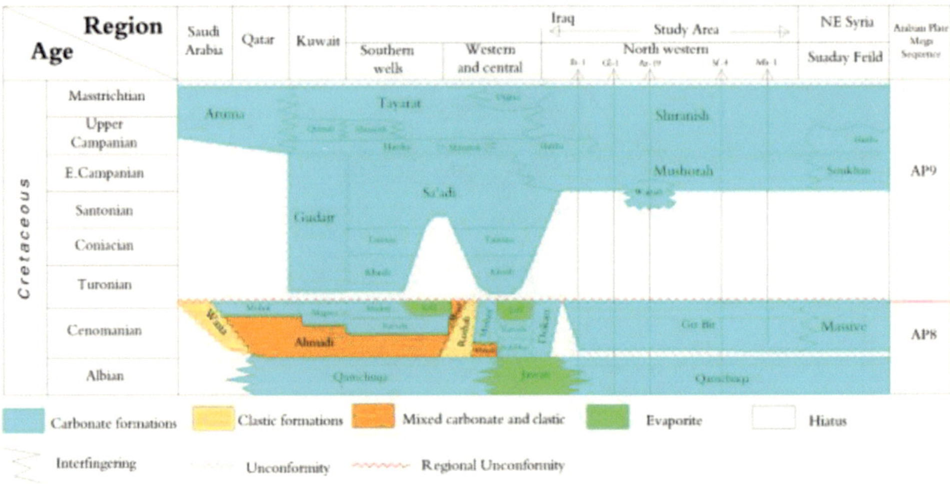

Figure 3.
Stratigraphic correlation of the cretaceous formations including the studied three formations in Iraq and neighboring countries [13].

4. Results

4.1 Lithologic and petrographic description

Lithologic characteristic of the carbonate succession from the three studied formations based on core samples including the commonly observed fractures and stylolites are conducted and representative samples are shown in **Figure 4**, whereas, representative microscopic images are presented in **Figure 5** to illustrate the main petrographic constituents.

4.1.1 Gir Bir formation

Lithologically, the formation dominantly is composed of carbonates in the form of recrystallized and dolomitized limestone with local silicification [16]. Petrographic investigation shows that the main skeletal components are benthonic foraminifera (Miliolids, Alveolina, and Orbitolina) and rudist while peloids and extraclasts are the main non-skeletal constituents (**Figure 5C** and **E**).

It is worth to mention that chemical compaction in the form of pressure solution or stylolitic textures is commonly observed in the studied carbonates of this formation. Carbonates of the Gir Bir Formation were suggested to be deposited in deep shelf margin with rudist buildups in a back reef/shoal environments [17].

4.1.2 Wajna formation

Carbonates of this formation are represented by dark gray limestones, locally silicified, alternated with thinly bedded dolomite including anhydrite nodules in the lower part of the formation and with white to gray limestone with common fractures and stylolites in the upper part of the formation [8, 16]. Petrographically, the carbonates of the formation is composed of benthonic foraminifera (Miliolids and Glomospira) ostracoda and peloids (**Figure 5A** and **B**) that were deposited in shallow marine environment that ranges from supratidal, protected lagoon to outer lagoonal (shoal) environments within two cycles in a regressive setting [8].

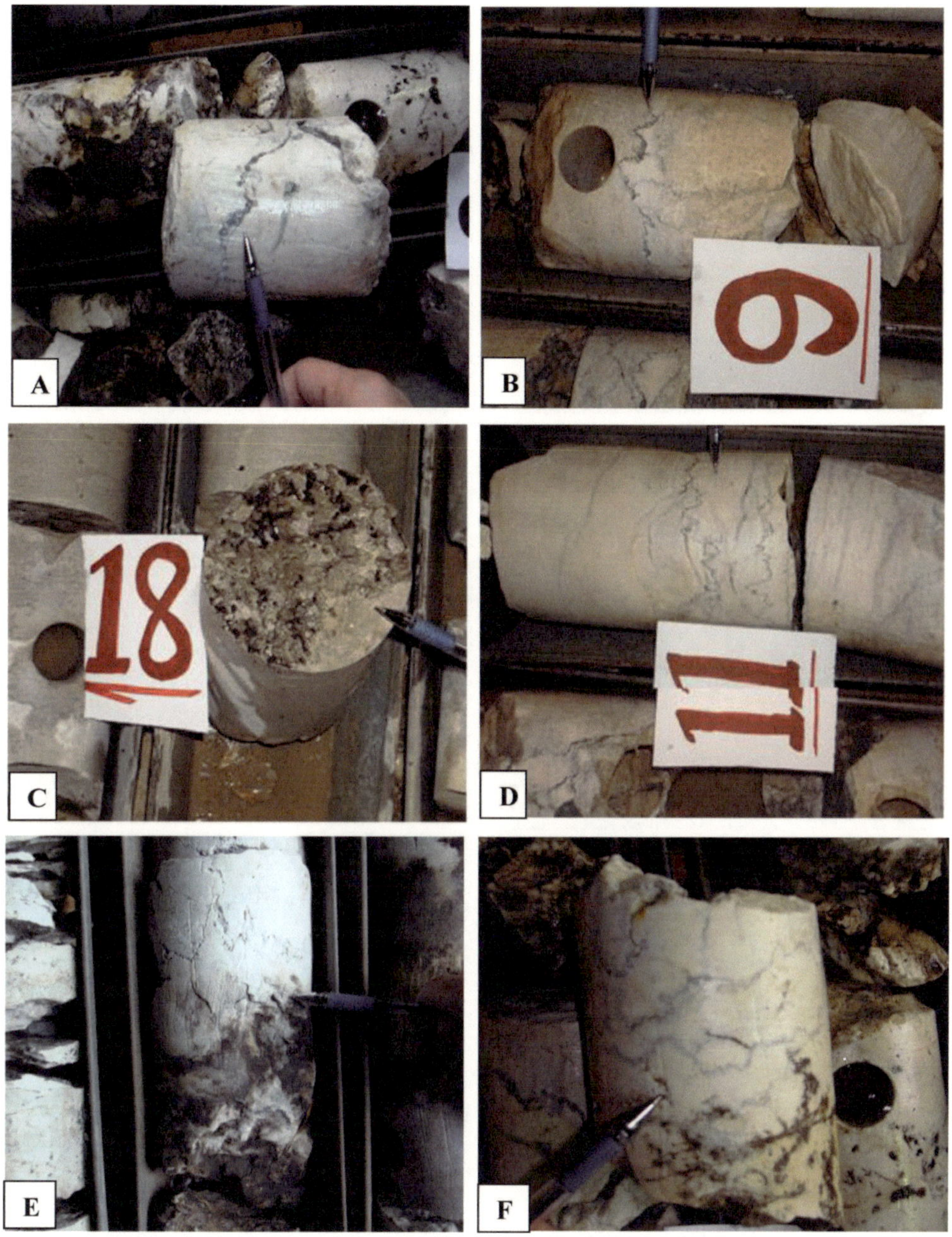

Figure 4.
Representative hand specimens of cores including commonly observed stylolites in the studied formations. A- Solution seams in the Gir Bir formation including organic matter and clayey materials. Well Bh-1, depth 1918 m. B- Stylolite from the Mushorah formation, well SF-16, depth 1844 m. C- Accumulation of organic matter at the lower surface of stylolite, Mushorah formation, well SF-16, depth 1816 m. D- Irregular anastomosing sets, in core from the Gir Bir formation. Well SF-16, depth 1850 m E- stop of hydrocarbon upward movement near the stylolite surfaces, Gir Bir formation well Bh-1, depth 1821 m F- Same as in E, Gir Bir formation well SF-16, depth 1848 m. See ***Figure 1*** *for well locations.*

4.1.3 Mushorah formation

The formation is composed of greenish-gray recrystallized, silicified and marly limestone (Bellen et al., [16]). Hard gray limestone with marl and shale units with common joints and stylolite dominate the lower part of the formation [8].

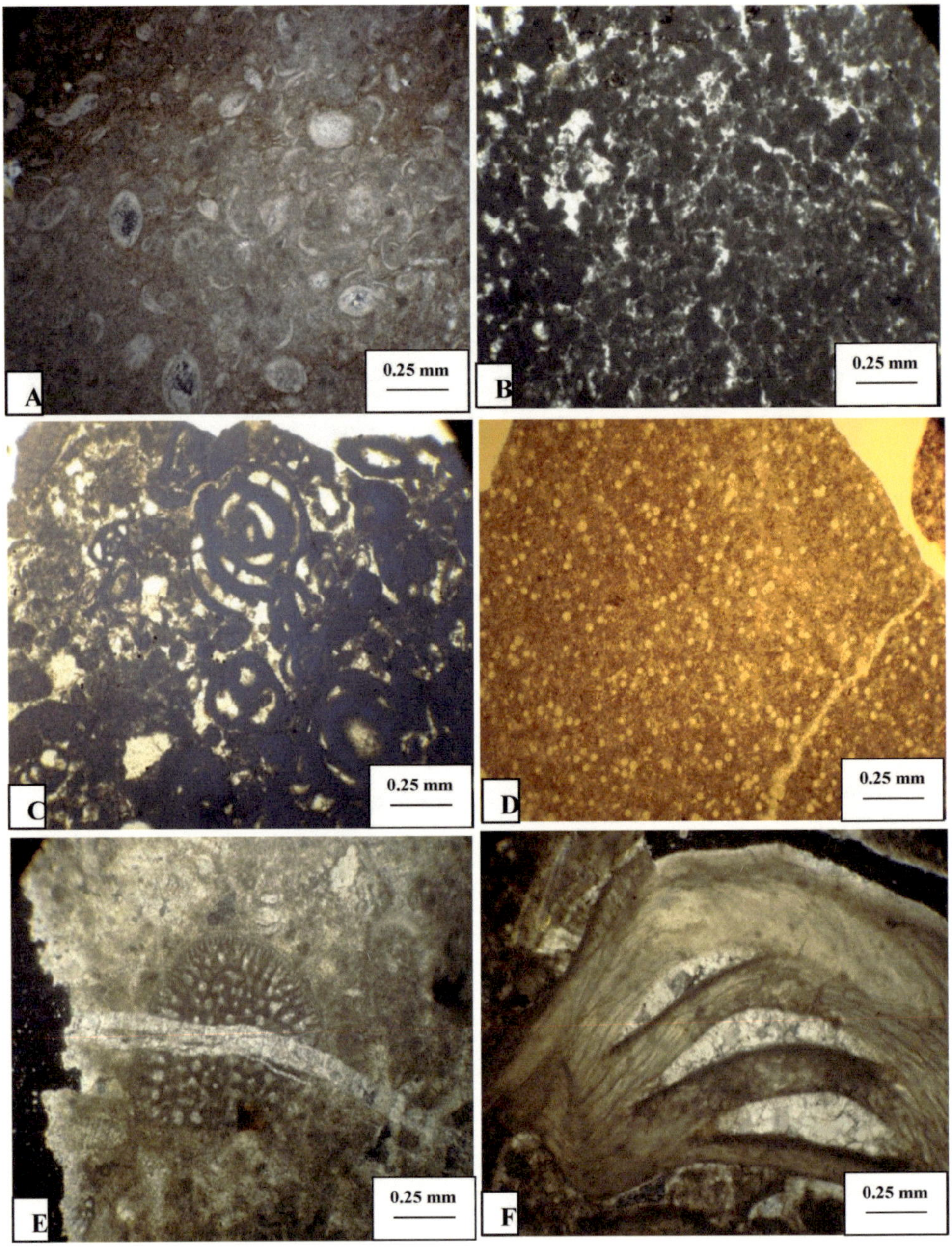

Figure 5.
Common petrographic constitutes in the studied formations. A- Ostracods from Wajna formation. B- Fecal pellets from Wajna formation. C- Milliolids from the Gir Bir formation. D- Calcispheres from Mushorah formation. E- Orbitolina foraminifera from Gir Bir formation, note microsparite filling veins crossing the foraminifera shell. F- Inoceramus from Mushorah formation.

Petrographic study shows that calcispheres, planktonic and benthonic foraminifera and Inoceramus form the main constituents of the formation (**Figure 5D** and **F**) that were deposited in a relatively deep marine environments ranging from outer shelf to upper bathyal.

The carbonates of the studied formation have been affected by several diagenetic processes such as; cementation, compaction, dolomitization, recrystallization, micritization, dissolution and silicification.

4.2 Pressure solution features

Two types of pressure solution features have been recognized in the current study; stylolites and solution seams based on variation in size and amplitude of the pressure solution surfaces in addition to thickness of the insoluble residues commonly exist in between these surfaces as a result of compaction on the carbonate rocks.

4.2.1 Stylolites

Stylolites are the main compaction features that have been recognized in the studied carbonates. They are defined as irregular surfaces formed due to different vertical movements under deep burial conditions [18]. They have various shapes such as columnar, pits, tooth-like etc., and can be commonly identified or differentiated by the concentration of insoluble residues that composed mainly of organic matter or clay minerals in between or near the stylolite surfaces.

In the present study, stylolites are identified either of large, macroscopic stylolites that easily recognized in the cores or finely-sized (microscopic) that are investigated using petrographic or scanning electron microscopes (**Figures 4**, **6** and 7).

Macroscopic stylolites in the core samples commonly are characterized by their lateral continuity and their varied thickness of their insoluble residues content. Most of these stylolites are horizontal or semi-horizontal and are parallel to bedding planes which may reflect their formation due to vertical movements [6].

According to classification of [19], several types of stylolites have been identified, these include;

1. Hummocky stylolite (**Figure 6A**), some of them are accompanied by dolomite rhombs (**Figure 6C**).

2. Irregular Anastomosing sets (**Figure 6B**)

3. Low-Amplitude peaks stylolites (**Figure 6F**)

4. High-Amplitude peaks stylolites (**Figure 6D**)

5. Irregular stylolites (**Figure 7A**)

Irregular stylolite is common in the Wajna and Mushorah formations while all types of stylolites are recorded in the Gir Bir Formation.

It is worth to mention that silicification is inhibited near the stylolite surfaces (**Figure 6E**).

High resolution imaging using scanning electron microscopy (**Figure 7B–F**) shows different forms of stylolites with concentration of authigenic minerals such as quartz and calcite near the stylolite surfaces.

4.2.2 Solution seams

They are low amplitude undulose surfaces with insoluble residues that may reach 1 centimeter thick. In the present work, low amplitude, laterally continuous solution seams with clay minerals as the main insoluble residues in between their surfaces are commonly seen (**Figure 4A**). Sometimes, a small-scale micro-stylolites accompanied these seams.

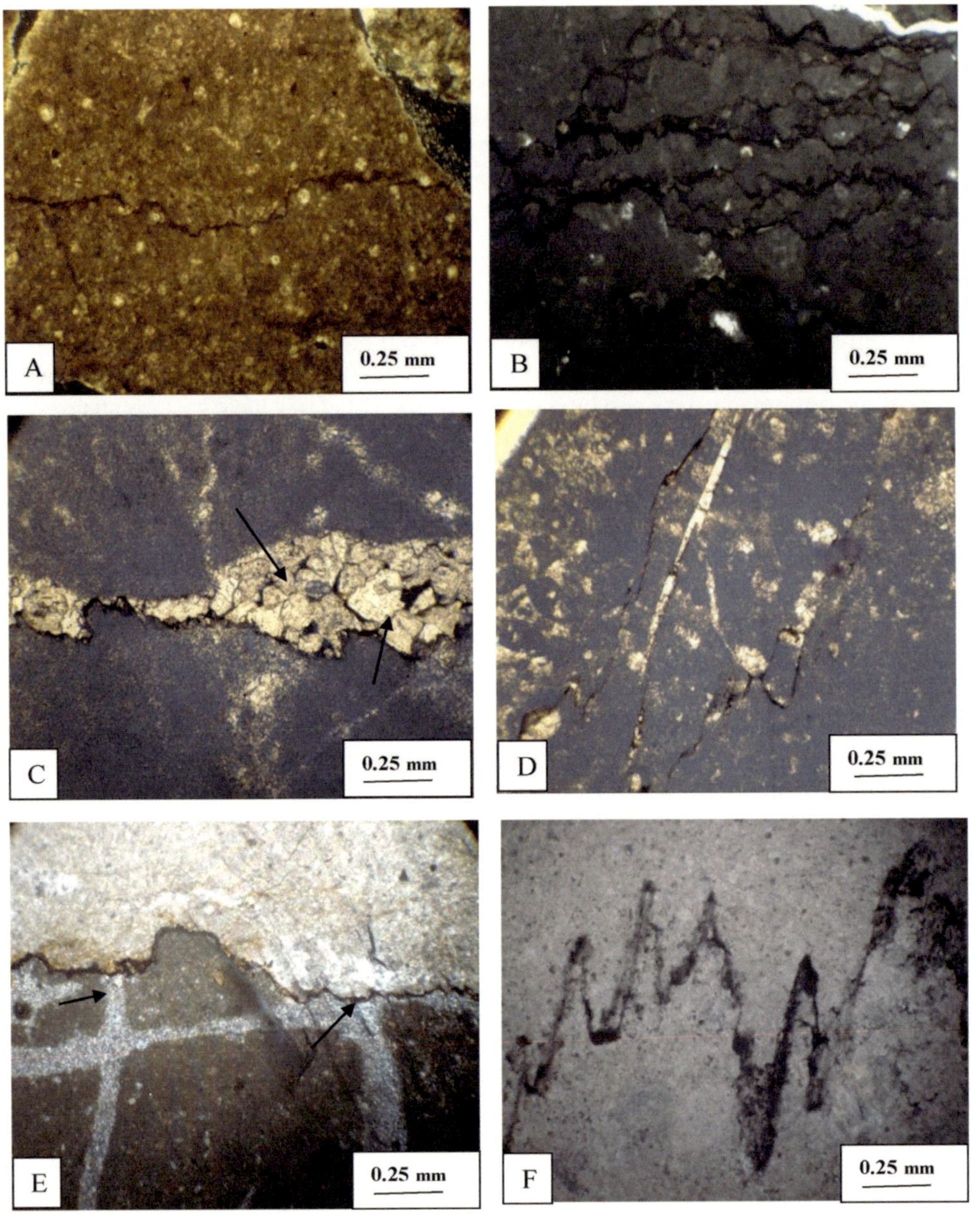

Figure 6.
A- Hummocky stylolite, Mushorah formation. B- Irregular anastomosing sets, Gir Bir formation. C- Dolomite rhombs (arrows) accompanied stylolite surfaces, Gir Bir formation. D- High-amplitude, Gir Bir formation. E- Stylolite inhibiting silicification (arrows), Mushorah formation. F- Low amplitude stylolites, Gir Bir formation.

5. Discussion

Pressure solution features in carbonate rocks are formed as a result of compaction. Stress increase pressure to critical level depending on rock composition then the rocks start to dissolve in order to relieve the stress [6]. This lead to decrease in the unit thickness which in turn affect the petrophysical characteristics (porosity and permeability) and thickness reduction compaction. Pores are commonly filled by cementing materials due to late diagenetic processes affecting on carbonate rocks [4, 20, 21]. Cementation is one of the common diagenetic processes affecting the

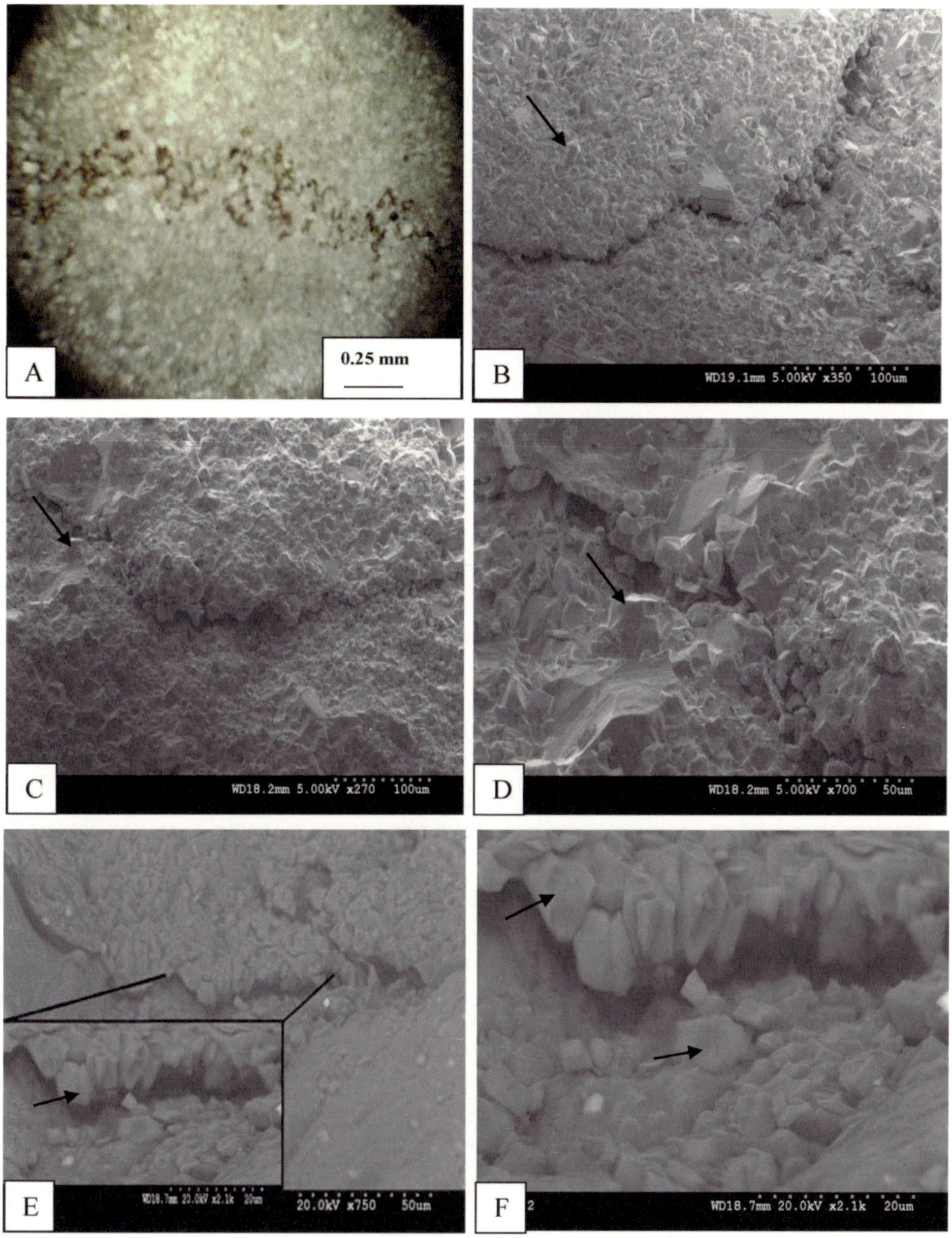

Figure 7.
A-Irregular stylolite, Gir Bir formation B-F, scanning electron micro images showing stylolites and authigenic mineral formation. B, C and D, stylolite with bypyramidal quartz (arrows), Mushorah formation. E, stylolite with calcite formation (arrow), F, enlarged view of E showing calcite grains (arrows).

studied carbonate succession. Various types of cements were recognized such as; blocky, fibrous, drusy mosaic and bladed [8]. Cementing materials inhibit vertical fluid movement which in turns prevent stylolite formation when they filled pores and increase the rock resistance to compaction.

Common occurrence of stylolites and solution seams in the upper part of the Gir Bir Formation (the main recorded reservoir in the present study) may be regarded as the main cause of hydrocarbon isolation and preventing them from upward migration as shown in core samples of the present study (**Figure 4E** and **F**). The main

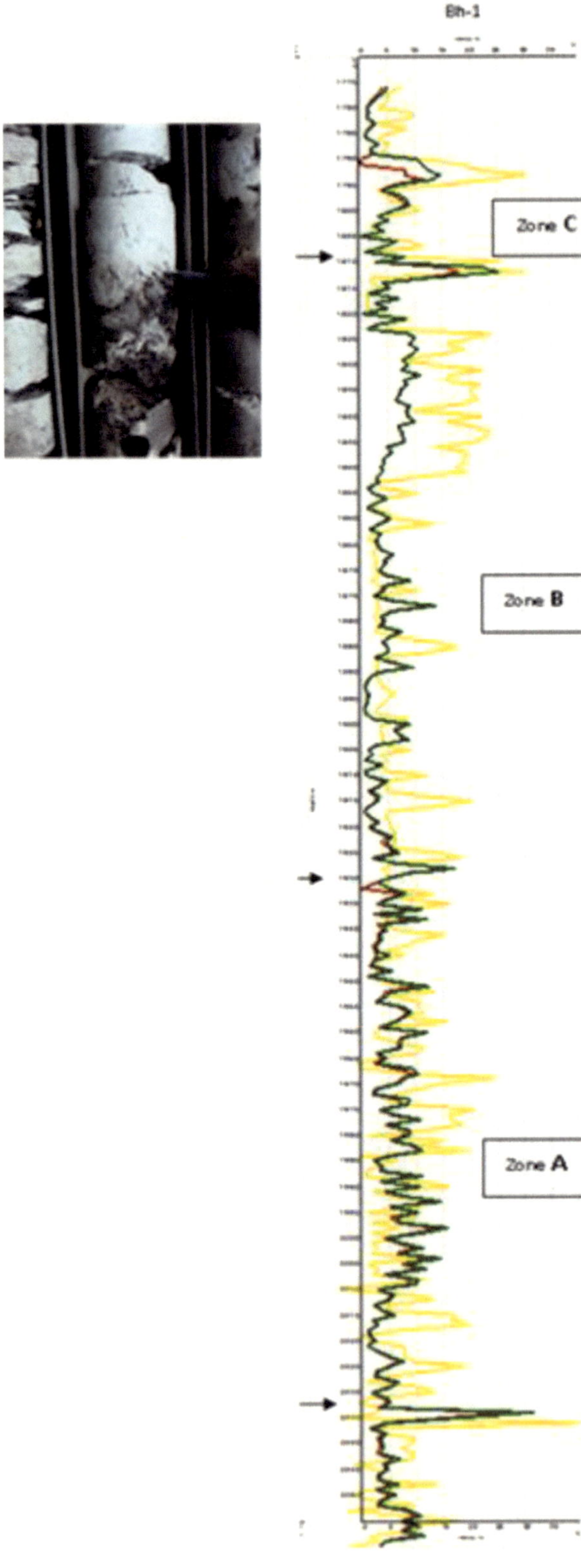

Figure 8.
Representative log showing the total porosity (green), effective porosity (red) and primary or original porosity (yellow) in Bh-1 well, note the higher porosity values in the upper part of zone B (Gir Bir formation) which declined abruptly in the area of rich stylolite occurrence and stop of upward hydrocarbon migration due to compaction and stylolite formation (core samples representing this location, depth 1821 m).

reason for that is the impermeable insoluble residues (organic matter and clay minerals) in between the stylolite surfaces that act as seals and stop the upward movement of hydrocarbons. Thickness of impermeable insoluble residues and lateral continuity of stylolite play a role in fluid movement [5].

Presence of hydrocarbons inhibit further formation of stylolite where the pores are filled with hydrocarbons that resist compaction and fluid movement and precipitation in the pores as a result of compaction [6].

There is also a close relationship between stylolites and minerals deposited by a percolating fluid. Scanning electron microscopic study revealed quartz deposited near the stylolite (**Figure 7B–D**) and calcite (**Figure 7E–F**) which may relate to presence of fluids saturated in Si or Ca that result in such mineralization [22]. Such processes also ended near the stylolite surfaces as seen in petrographic investigation (**Figure 6E**).

It is worth to mention that the Gir Bir Formation includes the studied successions were divided into three porosity zone (A, B and C) based on the values of porosity (primary, effective and total porosity) [8]. It is revealed that the Upper part of zone (B) which represent most of the upper part of the Gir Bir Formation is highly porous and regarded as the main reservoir in the studied succession.

After this high porosity values, the values declined directly in the area of high existence of stylolite in the Gir Bir Formation accompanied with end of hydrocarbon movement near the stylolite due to presence of impermeable materials in between stylolite surfaces (**Figure 8**).

6. Conclusions

Pressure solution features dominates in carbonate reservoir rocks of the Cenomanian-early Turonian Gir Bir Formation from northwestern Iraq which is regarded as the main reservoir in the upper Cretaceous succession of the area of study. Stylolites as common pressure solution results in reduction of unit thickness and porosity and permeability of the reservoirs. Pressure solution features act as a barrier for vertical movement of hydrocarbons in the studied successions especially in the upper part of the Gir Bir carbonate reservoir where the impermeable insoluble residues formed near the stylolite surfaces. These impermeable materials act as seals and stop the upward movement of hydrocarbons. Stylolites also act as a barrier and prevent further movement of mineralization processes such as silicification of carbonate (calcite or dolomite).

Acknowledgements

The authors gratefully acknowledge the Northern Oil Company, Kirkuk, Iraq for providing core samples. Thanks are due to Geology Department, University of Mosul, Iraq for support in petrographic analysis and to Mr. Georg Oleschinski from Steinmann Institute of Bonn University, Germany for SEM study.

Conflict of interest

All authors have participated in (1) conception and design, or analysis and interpretation of the data; (2) drafting the article or revising it critically for important intellectual content; and (3) approval of the final version.

The authors have no affiliation with any organization with a direct or indirect financial interest in the subject matter discussed in the manuscript.

Author details

Ali Al-Juboury[1*], Mohammed A. Al-Haj[1] and Aboosh H. Al-Hadidy[2]

1 Geology Department, University of Mosul, Mosul, Iraq

2 Imam Jaafar Al-Sadiq University, Kirkuk, Iraq

*Address all correspondence to: alialjubory@yahoo.com

References

[1] Al-Sakini JA. Summary of the petroleum geology of Iraq and the Middle East. Northern Oil Company Press (Naft –Al-Shamal co.), Kikuk, Iraq (in Arabic), 1992; p. 179

[2] Murris RJ. Middle East stratigraphic evolution and oil habitat. AAPG Bulletin. 1980; **64**:597-618

[3] Sharland PR, Archer R, Casey DM, Davies RB, Hall SH, Heward AB, Horbury AD, Simmons MD. Arabian Plate Sequence Stratigraphy, GeoArabia, Special Publication 2, Gulf Petrolink. Manama, Bahrain, 2001; p. 371

[4] Dunnington, HV. Aspects of diagenesis and shape change in stylolitic limestone reservoirs 7thWorld Petroleum Congress Mexico Proceedings, 1967; v2 p. 339

[5] Koepnick RB. Significance of stylolite development inhydrocar bon reservoirs with an emphasis on the Lower Cretaceous of the Middle East Geological Society of Malaysia Bulletin. 1988; **v22**, p. 23

[6] AlSharhan AS, Sadd JL. 2000. Stylolite in lower cretaceous carbonate reservoir. U.A.E. Middle East Models of Jurassic Cretaceous Carbonate Systems, SEPM Special Publication. 2000; **69**:185-207.

[7] Beydoun ZR. Arabian hydrocarbon geology and potential, plate tectonic approach. AAPG studies in Geology. 1991; 19-29

[8] Al-Hajj MA. Study of Sedimentary Basin Evolution of Cretaceous (Cenomanian-Early Campanian) Succession North Western Iraq, PhD thesis University of Mosul; 2011. p. 190

[9] Jassim SZ, Buday T. Late Tithonian-Early Turonian Megasequence AP8. In: Jassim SZ, Goff JC, (eds) "Geology of Iraq". Published by Dolin, Prague and Moravian Museum, Brno. 2006; 124-154

[10] Fouad SFA. Tectonic map of Iraq, scale 1: 1000 000, 3rd edition. Iraqi Bulletin of Geology and Mining, 2015; **11**: 1-7

[11] Buday T. The Regional Geology of Iraq, Stratigraphy and Paleogeography. Dar Al-Kutub publishing house, Mosul University. Mosul, Iraq, 1980; p. 445

[12] Dunnington HV. Generation, migration, accumulation and dissipation of oil in Northern Iraq. In: Weeks LG. (eds.) Habitat of Oil, A symposium. AAPG. Bull., 1958; 1149-1251

[13] Al-Naqib KM. Geology of the Arabian Peninsula, southern west Iraq". United States Geological survey, Professional Paper, 1967; No. 560-G, p. 54

[14] Haq BU, Hardenbol J, Vail BR. Mesozoic and Cenozoic Chrono-stratigraphy and Cycles of Sea-Level Change, In: Wilgus CK, Hastings BS, Kendall CG StC, Ross CA, Van Wagoner JC. Sea-Level Changes: An Integrated Approach: SEPM Special Publication. 1988; 42, SEPM, Tulsa, Oklahoma, USA.

[15] Haq BU, Al-Qahtani A. (2005) Phanerozoic cycles of sea-level change on the Arabian Platform. GeoArabia, Gulf petrolink, Bahrain, 2005; **10** (4): 127-160

[16] Bellen VRG, Dunnington HV, Wetzel R, Morton DM. Lexique Stratigraphic International". V.3, Asie Fascicule loa-Iraq. Paris, 1959; p. 333

[17] Al-Hajj MA, Al-Juboury AI, Al-Hadidy AH. Facies analysis and depositional environment of Gir Bir Formation, northwest Iraq, Scientific Journal of Sulaimania University, Kurdistan, Iraq, Special Issue, GeoKurdistan, 2016; **II**: 373-389.

[18] Toussaint R, Aharonov E, Koehn D, Gratier J-P, Ebner M, Baud P, Rolland A, Renardfh F. (2018) Stylolites: A review, Journal of Structural Geology, 2018; **114**: 163-195

[19] Logan BR, Semeniuk V. Dynamic metamorphism: Process and products in Devonian carbonate rocks. Carning basin, Western Australia, 1976; In Flugel E, (ed.) Microfacies Analysis of Limestone. Springer-Verlag Berlin, 1982; p. 633

[20] Wong PK, Oldershaw A. Burial cementation in the Devonian, Kaybob Reef Complex, Alberta, Canada, Journal of Sedimentary Research 1981; **51** (2): 507-520

[21] Oswald EJ, Muller HW, Goff DF, Al Habashi H, Al Matroushi S. Controls on porosity evolution in Thamama Group carbonate reservoirs in AbuDhabi, UAE, 9th Middle East Oil Show1995

[22] Park WC, Schot EH. Stylolites: their nature and origin. Journal of Sedimentary Petrology, 1968; **38**(1): 175-191

Chapter 3

Applications of Surfactants and Nanoparticles in Enhanced Oil Recovery Processes

Christian A. Paternina

Abstract

The surfactant injection is considered as the EOR (Enhanced Oil Recovery) with the highest potential to recover oil from reservoirs due to its ability to reduce interfacial forces into the porous medium. However, the adsorption of this type of chemical on the surface of rocks is the main problem when a surfactant injection project is applied since the surfactant molecules would rather be placed on rock minerals instead of being the oil–water interface. Based on this fact, this chapter would be discussed the significance of surfactant injection as an EOR method, the types of surfactants used, the main mechanism and parameters involved in the surfactant adsorption on the rock, and its consequences in oil recovery. Likewise, the addition of nanoparticles to inhibit the adsorption of surfactants is another topic that will be covered as a novel technology to improve the efficiency of the EOR process.

Keywords: Surfactants, Adsorption, EOR, Rock, Porous medium, nanoparticles

1. Introduction

In the hydrocarbon industry, there is a constant need to obtain large volumes of oil, which leads to the development of new techniques to extract the greater amount of oil into the reservoirs, which, using conventional technologies, would be impossible to achieve. The application of Enhanced Oil Recovery or EOR methods has been an opportunity to accomplish a substantial increase in hydrocarbon production in depleted reservoirs.

The surfactant injection is a recovery technique with great potential for recovery [1], which has shown an important effect in increasing the capillary number "Nc" (dimensionless variable, which relates viscous with interfacial forces). However, due to the great number of difficulties of its application in field, the efficiency of the process in many cases is very low. Among the most notable problems in the application of this method, it is the adsorption of surfactant on the rock, which leads to the chemical being trapped on the surface of the mineral substrate instead being positioned at the water–oil interface to reduce the interfacial tension.

2. Surfactant injection

Among the recovery methods, the surfactant injection has been listed as one with the greatest potential. It has been implemented in China, the United States,

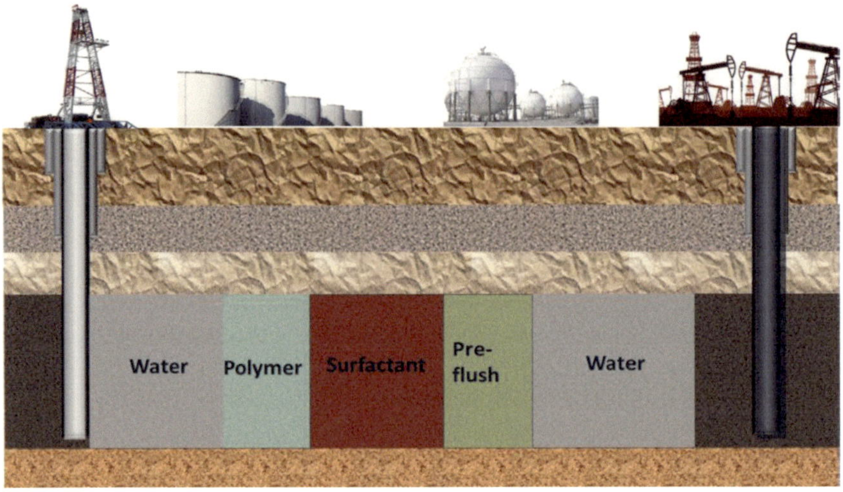

Figure 1.
Surfactant injection process.

France, Austria, Oman and Canada [2]; throughout history has achieved recovery factors up to 60% of the OOIP (Original Oil in Place) [3]. Due to the behavior of surfactants to reduce the interfacial tension at the oil–water interface to ultra-low values of up to 10^{-3} dynes/cm, easing the mobilization of the oil bank between 10% - 20% of the oil remaining in the formation, which was trapped in the porous medium. Successively, it can be pushed by a polymeric solution, which improves the mobility relationship between water and oil, favoring the final recovery factor achieved through the tertiary recovery technique [4].

A typical surfactant injection process is composed of different fluid stages or bumps, which can be distinguished in **Figure 1**. Thus, the first fluid injected into the formation consists of a fresh water pre-flush, which Its objective is to adjust or reduce the salinity of the formation, to avoid that the high salinity present in most oil fields, favors the precipitation or the adsorption of the surfactant to be injected; then, a surfactant formulation is displaced to transport the surfactants capable of positioning themselves between the oil–water phases and reducing the interfacial tension, favoring their movement to the surface. Afterwards, it is recommended to inject a bump of a high viscous polymer solution, intended to increase the viscosity of the aqueous solution and improve the sweep efficiency in the formation. Finally, water is injected to displace previously injected fluids.

2.1 Surfactant composition

Surfactants are amphiphilic molecules, which have a hydrophobic (non polar) part known as the "tail" and another hydrophilic (polar) called "head", as can be seen in **Figure 2**. The tail is made up of non polar groups (hydrocarbons and/or fluorocarbons) and the head, is composed of ions and/or polar compounds such as sulfates, sulfonates, carboxylate, phosphates or quaternary ammonium [5]. Surfactants due to their dual affinity nature, both for the hydrocarbon and the aqueous phase, can be used to improve recovery processes by reducing the free energy of the water–oil interface into reservoirs; resulting in an improved of the microscopic displacement efficiency [6].

The interfacial tension reduction achieved by surfactants is because of the adsorption of surfactant monomers in the interfacial region (water–oil) [7] and this fact makes the surfactant a versatile chemical which is widely used in different

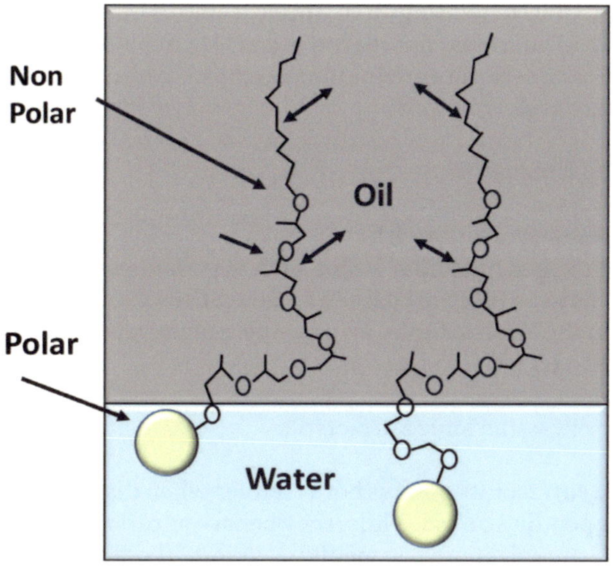

Figure 2.
Molecular structure of a surfactant.

kinds of processes in the oil industry such as: the formation of foams and emulsions, stabilization of fines, modification or change of wettability and of course enhanced oil recovery.

2.2 Types of surfactants

Surfactants are highly complex compounds used in a wide variety of industries such as pharmaceuticals, cosmetics, food, agrochemicals, and oil among others. In the hydrocarbon industry, surfactants have different uses among which can be mentioned: (de) emulsifiers, (anti) foaming agents, corrosion inhibitors, dispersants, humectants. Thus, there have been different forms to classify the surfactants, however, the most used what makes a distinction of the head nature head as shown below [8].

a. Anionic Surfactants:

Anionic surfactants have negatively charged hydrophilic group, consisting mainly of a sulfate, a sulfonate, a carboxylate or a phosphate. These surfactants dissociate into an amphiphilic ion and a cation, consisting of an alkali metal (Na^{+}, K^{+}) or a quaternary ammonium, as is the case of alkylbenzene sulfonate ($R\text{-}C_6H_4SO_3^{-}$ Na^{+}) [9]. Anionic surfactants are commonly used in enhanced oil recovery processes, since, thanks to its negative charge in the polar head, it is repelled by the clays and sandstones from the reservoirs, reducing its retention in the porous medium.

b. Cationic Surfactants:

Cationic surfactants have a positive charge on their polar head, usually made up of ammonium, pyridinium, or quaternary ammonium. When these surfactants are dissociated forming a cation and a halogen-type ion (Cl^{-}). One of the most used cationic surfactants is quaternary amine chloride

(R-N$(CH_3)_3^+Cl^-$). Regarding their applications in EOR, they are used mainly in carbonate formations, in order to reduce their adsorption, since the surface charge of limestone and dolomites is positive, and they repel surfactants of the same charge.

c. Non-ionic surfactants:

They are those surfactants with their polar part soluble in water, thanks to the inclusion of oxygen or similar atoms such as polyethylene glycol. However, they do not have an electric charge and when they dissociate in water, there is no change in the electric force, for example polyoxyethylated alkylphenol (R-C_6H_4 $(OC_2H_4)_X$-OH).

d. Amphoteric or Zwitterionic Surfactants:

Amphoteric surfactants possess both charges when dissociating in aqueous medium, depending on pH. Thus, they behave as cationic surfactants in acid solution and anionic in a basic medium. Among the chemical structures that are part of the polar head of the molecule are the compounds of amine oxide, betaine and carboxylated amines [10].

e. Gemini surfactants:

Gemini surfactants are a new family of surfactants that have at least two hydrocarbon tails and two polar or ionic groups, with a great variety of spacers of different nature, among which stand out short or long methylene groups, rigid groups (stilbene), polar (polyether) and nonpolar (aliphatic, aromatic) compounds. The ionic group can be positive (ammonium) or negative (phosphate, sulfate, carboxylate), while the polar non-ionic ones can be polyether or sugar [11].

The great advantage of Gemini surfactants over conventional surfactants is that they have a low CMC (Critical Micellar Concentration), high surface activity, better stability, high tolerance to hardness and can be used in low permeability reservoirs [12].

f. Biosurfactants:

Biosurfactants is a wide variety of amphiphilic molecules synthesized by plants, animal, and microbes. These molecules have a large range of molecular weight and are environmentally safe, making this kind of surfactants noteworthy for applications in oil industry and in the recent years in Enhanced Oil Recovery process specifically [13]. The use of biosurfactants have been applied in MEOR (Microbial in Enhanced Oil Recovery) since they have shown equal or better behavior than its chemical counterpart in several parameters especially for environmental compatibility [14]. However, this type of technology needs a large amount of resources and investment to overcome the traditional surfactants, even though the increase of green practices, there is a bright future for biosurfactant in EOR.

In **Figure 3** it is possible to observe the structure of the different types of surfactants according to their classification discussed above.

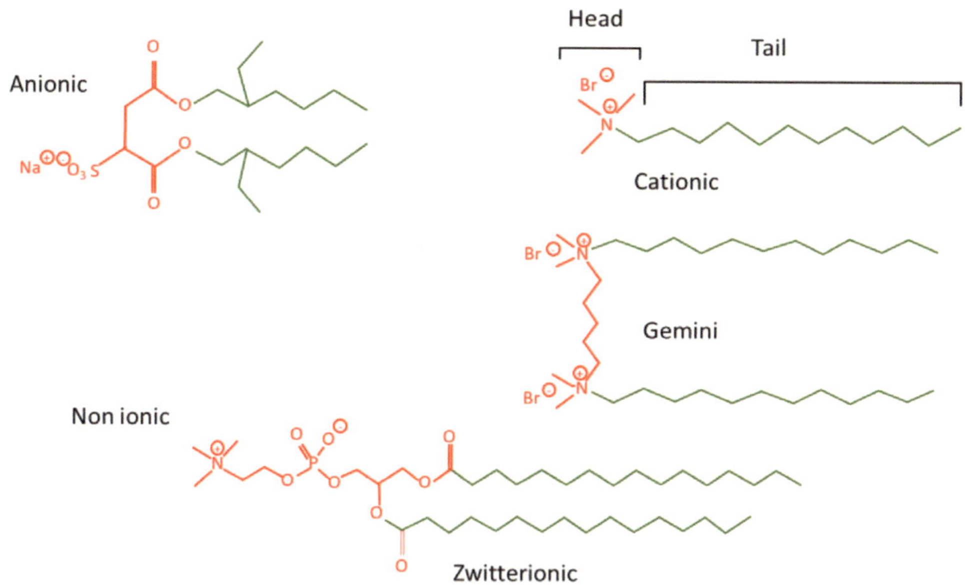

Figure 3.
Molecular structure of the different types of surfactants [15].

3. Adsorption of surfactants into the porous medium

Despite of the great potential shown by the surfactant injection for oil recovery, this methodology has high costs compared to other methods mainly due to the price of the injected fluids, additionally different problems arise when it comes to its field implementation in both technical and financial.

Among the many drawbacks of this recovery method, it can mention the loss of surfactants by their adsorption on the rocks of the reservoir, which has a large impact on determining whether the process is viable or not; since if the most of the injected chemical is trapped in the formation, it will not be available to locate at the oil–water interface and thus reduce the interfacial tension of the system [16, 17].

The adsorption of surfactants on a solid substrate can be defined as the selective distribution of the adsorbate around the surface of the rock medium, since it is energetically more favorable for the molecules than to be located at the oil–water interface [18]. The interactions of the surfactant molecules with the mineral substrate can be electrostatic or due to hydrophobic interactions as shown in **Figure 4.**

3.1 Surfactant adsorption mechanisms

To the surfactant being adsorbed at the solid–liquid interface, there are a series of mechanisms that cause the surfactant to adhere to the reservoir rock, such as:

a. Ion exchange:

Which involves the displacement or exchange of counterions adsorbed on the substrate by similarly charged ions. For example, a cationic surfactant can displace an adsorbed sodium cation on a clay and thus take its place on the surface of the mineral substrate (**Figure 5a**) [20–22].

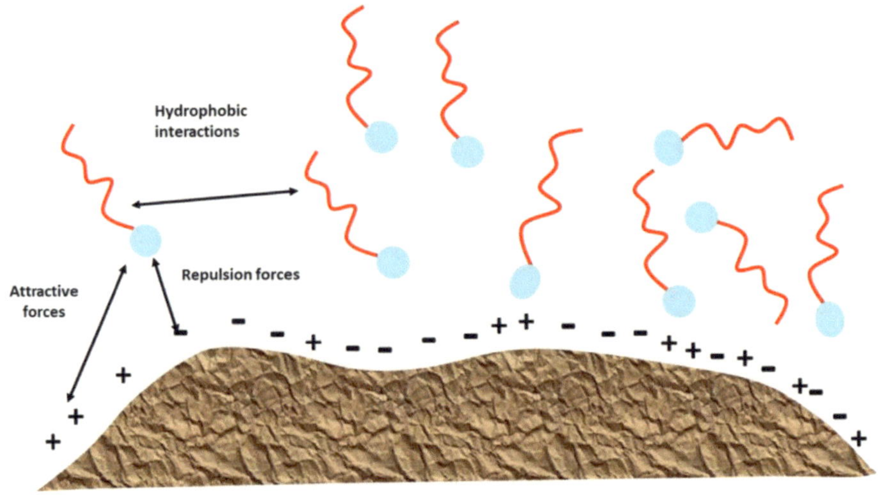

Figure 4.
Adsorption of surfactants on a substrate [19].

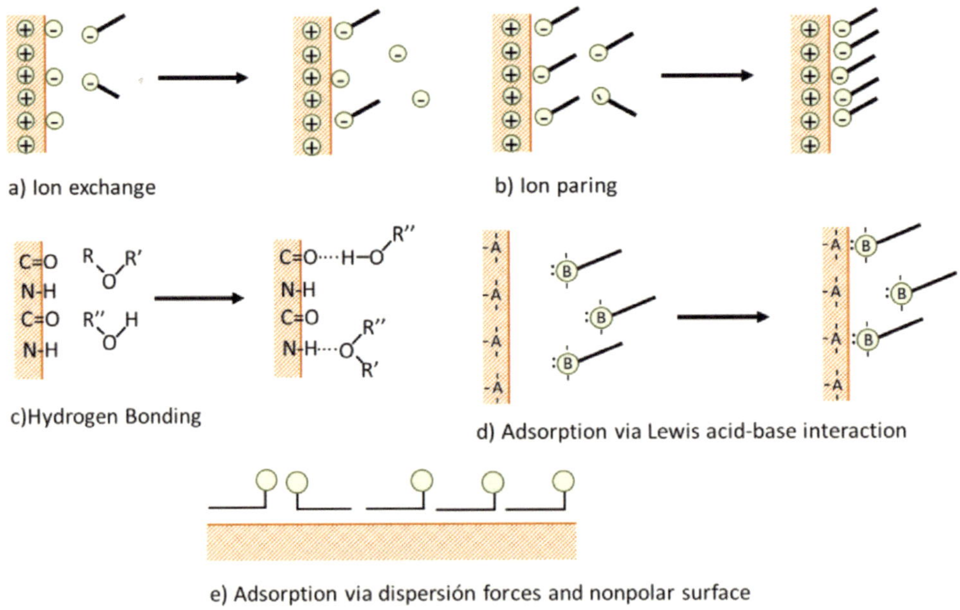

Figure 5.
Mechanisms of surfactant adsorption on solid substrates [5].

b. Ionic Pairing:

The adsorption of ionic surfactants on an oppositely charged solid surface to the surfactant polar head (**Figure 5b**) [21, 22].

c. Acid–Base Interactions:

Adsorption by means of hydrogen bonds (**Figure 5c**) between the substrate and the adsorbate (surfactant) [21–23] or by acid–base Lewis reaction, as seen in **Figure 5d**.

d. Adsorption by dispersion forces:

The London-Van der Waals dispersion forces act between the adsorbent molecules (mineral or solid substrate) and the adsorbate (surfactant) (**Figure 5e**). Adsorption by this type of mechanism generally increases with the molecular weight of the adsorbate and may be accompanied by other types of interactions that assist the adsorption, thus, this mechanism favors the ability of surfactant molecules to displace others from the interface [22, 24].

e. Hydrophobic bonds:

The combination of the attraction given by the hydrophobic groups of the surfactant molecules and their tendency to escape from the aqueous medium in which they are on allow the surfactants to be located on the solid surface due to the aggregation of their hydrocarbon chains [5].

3.2 Surfactant adsorption parameters

The adsorption of surfactants on the rocks from oil reservoirs is a phenomenon that is influenced by a wide number of parameters, however, there are some that produce a better synergy of the chemical with the rock surface, among which are:

a. Surfactant concentration:

It has been shown that, by increasing the concentration of surfactant in the aqueous solution in contact with a substrate, the amount of surfactant absorbed on said surface increases, since there is a greater availability of the chemical to be located in said spaces [2]. Thus, the importance to determinate the surfactant concentration to inject into a reservoir in EOR process, normally to values close to CMC (around 2000 ppm).

b. Temperature:

The increase in temperature leads to a slight decrease in the amount of surfactant adsorbed, due to a substantial increase in the translational kinetic energy and the entropy of the system, which reduces the attractive forces between the surfactant and the reservoir rock, avoiding the formation of an organized layer of surfactant monomers at the solid–liquid interface [2].

c. Hydrogen potential, pH:

The pH of the aqueous medium essentially influences the adsorption, because rock surfaces are positively or negatively charged, depending on the degree of dissociation of functional groups on their surface, so a change in pH alters the surface charge of minerals [25]. Thus, the amount of surfactant adsorbed on a solid can be altered by changing the pH of the medium. Therefore, in general, in the case of anionic surfactants (with a negative charge), adsorption decreases at low pH and for cationic surfactants (positive charge) it increases with pH [26].

d. Salinity:

The presence of salts in the reservoir affects the solubility and the formation of more complex aggregates formed by the surfactant molecules [27] due to the

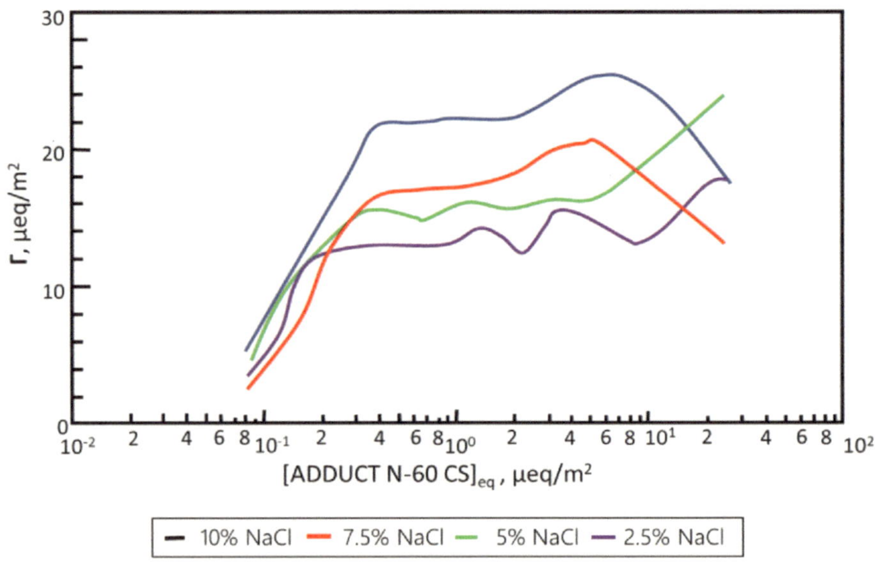

Figure 6.
Effect of NaCl concentration on the adsorption of an anionic surfactant [34].

electrostatic interactions of "salting-in" and "salting-out.", Where salts act as facilitators of aggregate formation [28, 29]. Additionally, when a compression of the so-called electrical double layer occurs [30, 31], the adsorption density of ionic surfactants at the interface is modified and the structure and morphology of the layers of micellar aggregates is changed [32]. On the other hand, the presence of electrolytes decreases the CMC (Critical Micellar Concentration) since it produce the decreasing of the repulsive forces between the surfactant molecules [19, 33].

Figure 6, it is possible to see the study case developed by Figdore [34], where it is described the adsorption of an anionic surfactant on a sandstone substrate at different NaCl concentrations in the aqueous medium [34]. Adsorption isotherms are observed, where it is possible to observe that meanwhile salt concentration in the medium increases, the adsorption of the surfactant also increases proportionally. This is attributed to the reduction of the repulsive forces between the negative heads of the surfactants and the surface of the solid substrate (also negatively charged), due to the electrolytes (Na +), allowing the surfactant molecules to adhere with easier on the surface of the sand grains.

a. Mineralogy:

It is of great importance to identify the minerals that make up the solid substrate or porous medium, which will be contacted by the injected surfactants since depending on which components form the surface of the rock, certain adsorption mechanisms exist and will give a guideline for thus select the type of surfactants that should be implemented in an enhanced recovery processes to reduce the chemical adsorption on the rock. Typically, surfactants with the same charge to the substrate of interest are selected to reduce the magnitude of adsorption [35].

Examples of common mineralogies in oil and gas reservoirs are quartz sandstones and clay minerals such as kaolinite.

- Kaolinite:

It is formed by a clay mineral with the chemical formula $Al_2 (OH)_4Si_2O_5$, which is a claystone, with a laminar structure, made up of a tetrahedral sheet of silica and another, octahedral of alumina [7]. At neutral pH, kaolinite is negatively charged on the faces of its surface and positively on the edges of the same. The isoelectric point of kaolinite occurs at low pH, but it is at pH values close to or greater than 4.6, the negative charge density increases significantly on its surface [36]. Additionally, the level of adsorption of surfactants on kaolinite is higher compared to other porous solids since it has a large surface area [37].

- Sandstone:

It is a type of rock, in which silica prevails. This compound acquires a surface charge depending on the relative concentration of H^+ and OH^- in solution; Therefore, the charge of this surface depends mainly on the pH. On the other hand, the isoelectric point of silica occurs approximately at a pH of 2, showing that the negative charge density remains low, until the pH reaches values higher than 6 [38].

Other parameter that has to be kept on mind is the use of the thermodynamic potential of the Gibbs free energy, which relates the variations of the enthalpy and entropy of a system, indicating the spontaneity of the reaction and in our case of the adsorption. Thus, the greater the negative of the Gibbs energy variation, the process will be energetically favored.

4. Influence of nanoparticles on surfactant adsorption

Due to the inconveniences experienced by the injection of surfactants, in which there are a low reduction in interfacial tension, high technical costs and large amounts of surfactant adsorbed on the rock; new technologies have been proposed to optimize surfactant injection processes, where the application of nanoparticles is seen as a proposal to improve the technique performance.

Broadly speaking, the application of nanoparticles in EOR processes can be divided into three main types: nanofluids, nanoemulsions and nanocatalysts. The first two have great applicability in surfactant recovery techniques since they seek to influence capillary forces to obtain an increase in the recovery factor. The EOR mechanisms of nanofluids have already been investigated in literatures, which mainly includes disjoining pressure, pore channels plugging, viscosity increase of injection fluids, IFT reduction, wettability alteration and preventing asphaltene precipitation. On the other hand, the use of nanocatalysts has shown great applicability in thermal recovery processes, since nanoparticles allow an accelerated decrease in the viscosity of crude oil [39]. However, the level of viscosity reduction is related to the types of nanoparticles, their concentration, and fluid temperature.

There are different studies that affirm that the addition of nanoparticles to a solution with surfactants can be beneficial for increasing the recovery factor [40–42]. However, the predominant mechanisms that favor the increase in oil recovery have not been determined with certainty. Reduction on the adsorption of surfactants on the rock surface could be a powerful mechanism to improve recovery processes. However, the number of studies of this type has been limited. However, in the existing studies of this topic, it has been shown that nanoparticles produce a reduction in the amount of surfactant adsorbed on the rock [43–48] as can be seen in **Figure** 7, meanwhile the nanoparticle of Silica oxide concentration increases the amount of adsorbed SDS (Sodium dodecyl sulfate) surfactant decreases.

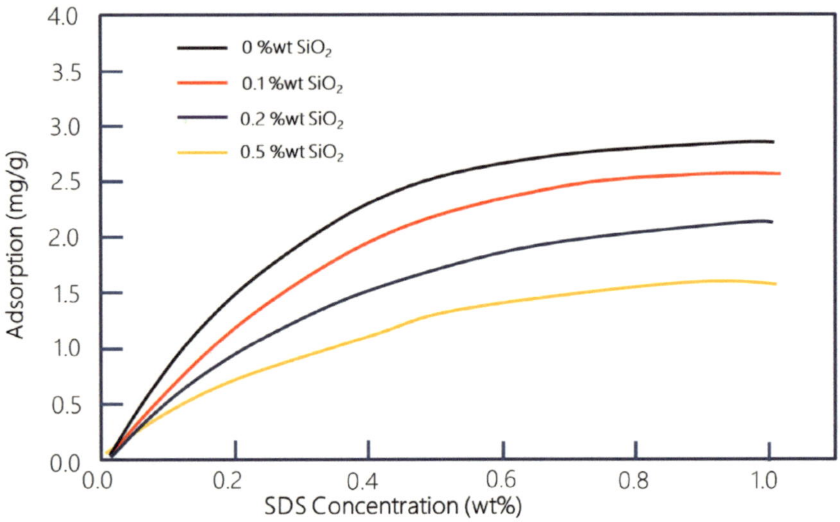

Figure 7.
Effect of the addition of silica oxide nanoparticles on the adsorption of SDS surfactant [45].

In the Zargartalebi study of 2014, two types of silica nanoparticles of different hydrophobicity were implemented with an anionic surfactant SDS (sodium Dodecyl Sulfate) on a substrate to determine the effect on the adsorption of the surfactant. The results of the experiment showed that indeed in both cases there was a reduction in the magnitude of the adsorption in general at lower values of the CMC of the surfactant used. It was observed that the hydrophobic nanoparticle showed a greater reduction in the amount of surfactant adsorbed due to the amphoteric behavior of this type of nanoparticle, which allowed the nanoparticles to be more easily located on the substrate than the hydrophilic ones, preventing the surfactant from being trapped on the rock surface [43].

Even so, the mechanisms that cause the reduction in the adsorption of surfactants when using nanoparticles have not been fully elucidated either, although several theories have been raised. For example, that negatively charged nanoparticles

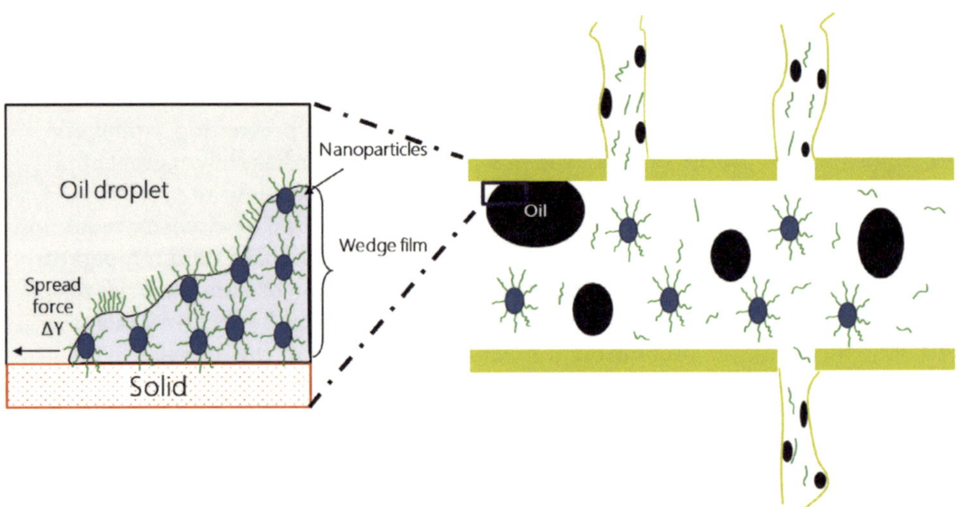

Figure 8.
Synergistic combination of surfactants and nanoparticles for EOR processes [51].

and anionic surfactant molecules compete for adsorption on the substrate surface due to electrostatic interactions of charged compounds with the solid surface [45]; for this reason, at low concentrations of surfactant this competition for being located on the surface prevents a large amount of surfactant from being located on the surface. However, by increasing the concentration of surfactant above the concentration of nanoparticles, the surfactant has a greater preference to position itself on the surface of the substrate [49]. Another theory states that the inhibition of surfactant adsorption is due to the formation of aggregates between the surfactants and the nanoparticles which tend to remain in the aqueous solution, also due to the negative charge of both the aggregates and the substrate, the repulsive forces keep the surfactants stable in the solution and prevent them from adsorbs on the surface of the substrate [50] as observed in **Figure 8**, where it is noted that the surfactant molecules adhere to the nanoparticles and prevent interactions with the substrate of the pore throats into reservoirs [51].

5. Nanoparticles influence in IFT and wettability

Additionally, of adsorption, the change of IFT and wettability are the most important EOR mechanisms reported for surfactant with nanoparticles process. Although the silica nanoparticles have demonstrated a good behavior in these topics, the use of another nanoparticle type have been more promising.

Various researches have been focused on the assisting of SDS (Sodium Dodecyl Sulfate) surfactant with other type of nanoparticle [48, 52, 53]. These studies showed that the incorporation of ZrO2 nanoparticles allows to displace more easily, the crude trapped in the porous medium [54], since the IFT values are drastically reduced, showing the efficiency of the ZrO2/SDS solution to reduce capillary pressure in the poral space, which allows increasing the mobilization of the crude oil [52]. Furthermore, the presence of nanoparticles can change rheological properties and increase the effectiveness of the solution in the recovery process [54].

Aluminum oxide nanoparticles also have demonstrated good performance combined with anionic surfactants displaying the effectiveness of aluminum-based nanofluids in altering the wettability of sandstone cores [55]. The results revealed that the surfactant acts by modifying the surface and that its effect can be enhanced by adding Al2O3 nanoparticles at low concentrations.

6. Conclusions

The surfactant injection is a promising EOR method that is in charge to reduce the interfacial tension of the water–oil interface to recover an additional amount of petroleum into the reservoir, however, a large part of the surfactant compound is entrapment on the rock surface due to interactions of the chemical and the minerals causing surfactant loss and low recovery factors.

The surfactant adsorption on the rock is a phenomenon that depends on various parameters like salinity, mineralogy, temperature, and pH which depending on the magnitude of these ones, the type of interaction of rock and surfactant will be determined. Likewise, the range of adsorption will be decided by the type of adsorption mechanism and parameters present.

The use of new techniques to reduce surfactant adsorption has been used to increase the efficiency of the recovery factor. Among the most popular, the use of nanoparticles has shown an excellent performance since reduce the adsorption of surfactant to interact with the rock surface and the surfactant as well.

Acknowledgements

I express my gratitude to the UIS (Universidad Industrial de Santander) for let me develop my researches focus on adsorption phenomena in its facilities.

Conflict of interest

The author declares no conflict of interest.

Author details

Christian A. Paternina
Universidad Industrial de Santander, Bucaramanga, Colombia

*Address all correspondence to: christian.paternina@correo.uis.edu.co

References

[1] Green DW, Willhite GP. Enhanced oil recovery. Vol. 6. Henry L. Doherty Memorial Fund of AIME, Society of Petroleum Engineers; 1998.

[2] Dang CTQ, Chen ZJ, Nguyen NTB, Bae W, Phung TH. Development of Isotherm Polymer/Surfactant Adsorption Models in Chemical Flooding. In Society of Petroleum Engineers; 2011 [cited 2020 Feb 22]. Available from: https://onepetro.org/conference-paper/SPE-147872-MS

[3] Emegwalu CC. Enhanced oil recovery for Norne Field's E-segment using surfactant flooding. Norges teknisk-naturvitenskapelige universitet, Fakultet for ...; 2010.

[4] Chávez G. Reparto Preferencial de Alcoholes Polietoxilados. [Maracaibo]: Universidad de Zulia; 2006.

[5] Rosen MJ, Kunjappu JT. Surfactants and interfacial phenomena. Vol. 82. Wiley Online Library; 2004.

[6] Romero-Zeron L. Advances in Enhanced Oil Recovery Processes, University of New Brunswick. Chem Eng Dep Can. 2012.

[7] Schramm LL. Surfactants: fundamentals and applications in the petroleum industry. New York: Cambridge University Press; 2000.

[8] Myers D. Surfactant science and technology. Wiley Online Library; 2006.

[9] Salager J. Surfactantes-Tipos y Usos (Cuaderno FIRP S300-A). Módulo Enseñ En Fenóm Interfaciales Univ Los Andes. 2002;3:1-54.

[10] Farn RJ, editor. Chemistry and Technology of Surfactants. Edición: 1. Oxford ; Ames, Iowa: Wiley-Blackwell; 2006. 336 p.

[11] Menger FM, Littau CA. Gemini surfactants: a new class of self-assembling molecules. J Am Chem Soc. 1993 Nov 1;115(22):10083-10090.

[12] Chauhan PD. Data analysis and summary for surfactant-polymer flooding based on oil field projects and laboratory data. 2014.

[13] Jahan R, Bodratti AM, Tsianou M, Alexandridis P. Biosurfactants, natural alternatives to synthetic surfactants: Physicochemical properties and applications. Adv Colloid Interface Sci. 2020 Jan 1;275:102061.

[14] S.j. G, Banat IM, Joshi SJ. Biosurfactants: Production and potential applications in microbial enhanced oil recovery (MEOR). Biocatal Agric Biotechnol. 2018 Apr 1;14:23-32.

[15] Eastoe J, Tabor RF. Surfactants and nanoscience, colloidal foundations of nanoscience. Colloidal Found Nanosci Elsevier Amst. 2014;

[16] Ahmadall T, Gonzalez MV, Harwell JH, Scamehorn JF. Reducing Surfactant Adsorption in Carbonate Reservoirs. SPE-24105-PA. 1993 May 1;8(02):117-22.

[17] Thomas S. Enhanced oil recovery-an overview. Oil Gas Sci Technol-Rev IFP. 2008;63(1):9-19.

[18] Somasundaran P, Huang L. Adsorption/aggregation of surfactants and their mixtures at solid–liquid interfaces. Adv Colloid Interface Sci. 2000 Dec 11;88(1):179-208.

[19] Ahmadi MA, Shadizadeh SR. Experimental investigation of a natural surfactant adsorption on shale-sandstone reservoir rocks: Static and dynamic conditions. Fuel. 2015 Nov 1;159:15-26.

[20] WAKAMATSU T, FUERSTENAU DW. The Effect of Hydrocarbon Chain Length on the Adsorption of Sulfonates at the Solid/Water Interface. In: Adsorption From Aqueous Solution [Internet]. AMERICAN CHEMICAL SOCIETY; 1968 [cited 2020 Apr 19]. p. 161-72. (Advances in Chemistry; vol. 79). Available from: https://doi.org/10.1021/ba-1968-0079.ch013

[21] Rupprecht H, Liebl H. INFLUENCE OF TENSIDES ON COLLOID CHEMICAL BEHAVIOR OF HIGHLY DISPERSE SILICIC ACIDS IN POLAR AND NONPOLAR SOLVENTS. KOLLOID-Z Z Polym. 1972;250(7):719–+.

[22] Law Jr JP, Kunze GW. Reactions of surfactants with montmorillonite: Adsorption mechanisms. Soil Sci Soc Am J. 1966;30(3):321-327.

[23] Snyder LR. Interactions responsible for the selective adsorption of nonionic organic compounds on alumina. Comparisons with adsorption on silica. J Phys Chem. 1968;72(2):489-494.

[24] Kölbel H, Kühn P. Konstitution und Eigenschaften grenzflächenaktiver Stoffe I. Messungen an p-n-Alkyl-Benzolsulfonaten. Angew Chem. 1959;71(6):211-215.

[25] Lv W, Bazin B, Ma D, Liu Q, Han D, Wu K. Static and dynamic adsorption of anionic and amphoteric surfactants with and without the presence of alkali. J Pet Sci Eng. 2011 May 1;77(2):209-218.

[26] Solairaj S, Britton C, Kim DH, Weerasooriya U, Pope GA. Measurement and Analysis of Surfactant Retention. In Society of Petroleum Engineers; 2012 [cited 2019 May 7]. Available from: https://www.onepetro.org/conference-paper/SPE-154247-MS

[27] Rico-Rico Á, Temara A, Behrends T, Hermens JLM. Effect of sediment properties on the sorption of C12-2-LAS in marine and estuarine sediments. Environ Pollut. 2009 Feb 1;157(2):377-383.

[28] Fernández-Castro B, Méndez-Morales T, Carrete J, Fazer E, Cabeza O, Rodríguez JR, et al. Surfactant Self-Assembly Nanostructures in Protic Ionic Liquids. J Phys Chem B. 2011 Jun 30;115(25):8145-8154.

[29] Bickel KR, Timm AE, Nattland D, Schuster R. Microcalorimetric Determination of the Entropy Change upon the Electrochemically Driven Surface Aggregation of Dodecyl Sulfate. Langmuir. 2014 Aug 5;30(30):9085-9090.

[30] Azam MR, Tan IM, Ismail L, Mushtaq M, Nadeem M, Sagir M. Static adsorption of anionic surfactant onto crushed Berea sandstone. J Pet Explor Prod Technol. 2013 Sep 1;3(3):195-201.

[31] Tay A, Oukhemanou F, Wartenberg N, Moreau P, Guillon V, Delbos A, et al. Adsorption inhibitors: a new route to mitigate adsorption in chemical enhanced oil recovery. In: SPE Asia Pacific enhanced oil recovery conference. Society of Petroleum Engineers; 2015.

[32] Chang Z, Chen X, Peng Y. The adsorption behavior of surfactants on mineral surfaces in the presence of electrolytes – A critical review. Miner Eng. 2018 Jun 1;121:66-76.

[33] Wennerstrom H, Khan A, Lindman B. Ionic surfactants with divalent counterions. Adv Colloid Interface Sci. 1991 Jan 1;34:433-449.

[34] Figdore PE. Adsorption of surfactants on kaolinite: NaCl versus CaCl2 salt effects. J Colloid Interface Sci. 1982 Jun 1;87(2):500-517.

[35] Nevskaia DM, Guerrero-Ruíz A, de D. López-González J. Adsorption of

Polyoxyethylenic Nonionic and Anionic Surfactants from Aqueous Solution: Effects Induced by the Addition of NaCl and CaCl2. J Colloid Interface Sci. 1998 Sep 1;205(1):97-105.

[36] Schroth BK, Sposito G. Surface Charge Properties of Kaolinite. Clays Clay Miner. 1997 Feb 1;45(1):85-91.

[37] Amirianshoja T, Junin R, Idris AK, Rahmani O. A comparative study of surfactant adsorption by clay minerals. J Pet Sci Eng. 2013;101:21-27.

[38] Atkin R, Craig VSJ, Wanless EJ, Biggs S. Mechanism of cationic surfactant adsorption at the solid–aqueous interface. Adv Colloid Interface Sci. 2003 May 30;103(3):219-304.

[39] Almahfood M, Bai B. The synergistic effects of nanoparticle-surfactant nanofluids in EOR applications. J Pet Sci Eng. 2018 Dec 1;171:196-210.

[40] Suleimanov BA, Ismailov FS, Veliyev EF. Nanofluid for enhanced oil recovery. J Pet Sci Eng. 2011 Aug 1;78(2):431-437.

[41] Roustaei A, Moghadasi J, Bagherzadeh H, Shahrabadi A. An Experimental Investigation of Polysilicon Nanoparticles' Recovery Efficiencies through Changes in Interfacial Tension and Wettability Alteration. In Society of Petroleum Engineers; 2012 [cited 2018 Nov 28]. Available from: https://www.onepetro.org/conference-paper/SPE-156976-MS

[42] Hendraningrat L, Torsaeter O. Unlocking the Potential of Metal Oxides Nanoparticles to Enhance the Oil Recovery. In Offshore Technology Conference; 2014 [cited 2018 Nov 28]. Available from: https://www.onepetro.org/conference-paper/OTC-24696-MS

[43] Zargartalebi M, Barati N, Kharrat R. Influences of hydrophilic and hydrophobic silica nanoparticles on anionic surfactant properties: Interfacial and adsorption behaviors. J Pet Sci Eng. 2014 Jul 1;119:36-43.

[44] Zargartalebi M, Kharrat R, Barati N. Enhancement of surfactant flooding performance by the use of silica nanoparticles. Fuel. 2015 Mar 1;143:21-27.

[45] Wu Y, Chen W, Dai C, Huang Y, Li H, Zhao M, et al. Reducing surfactant adsorption on rock by silica nanoparticles for enhanced oil recovery. J Pet Sci Eng. 2017 May 1;153:283-287.

[46] Saxena N, Kumar A, Mandal A. Adsorption analysis of natural anionic surfactant for enhanced oil recovery: The role of mineralogy, salinity, alkalinity and nanoparticles. J Pet Sci Eng. 2019 Feb 1;173:1264-1283.

[47] Yekeen N, Manan MA, Idris AK, Samin AM, Risal AR. Experimental investigation of minimization in surfactant adsorption and improvement in surfactant-foam stability in presence of silicon dioxide and aluminum oxide nanoparticles. J Pet Sci Eng. 2017 Nov 1;159:115-134.

[48] Yekeen N, Padmanabhan E, Idris AK, Ibad SM. Surfactant adsorption behaviors onto shale from Malaysian formations: Influence of silicon dioxide nanoparticles, surfactant type, temperature, salinity and shale lithology. J Pet Sci Eng. 2019 Aug 1;179:841-854.

[49] Paria S, Khilar KC. A review on experimental studies of surfactant adsorption at the hydrophilic solid–water interface. Adv Colloid Interface Sci. 2004;110(3):75-95.

[50] Esmaeilzadeh P, Fakhroueian Z, Bahramian A, Arya S. Influence of ZrO2 Nanoparticles including SDS and CTAB Surfactants Assembly on the Interfacial Properties of Liquid-Liquid, Liquid-Air

and Liquid-Solid Surface Layers. In: Journal of Nano Research. Trans Tech Publ; 2013. p. 15-21.

[51] Zhao M, Lv W, Li Y, Dai C, Wang X, Zhou H, et al. Study on the synergy between silica nanoparticles and surfactants for enhanced oil recovery during spontaneous imbibition. J Mol Liq. 2018 Jul 1;261:373-378.

[52] Moslan MS, Sulaiman WRW, Ismail AR, Jaafar MZ. Applications of Aluminium Oxide and Zirconium Oxide Nanoparticles in Altering Dolomite Rock Wettability using Different Dispersing Medium. Chem Eng Trans. 2017 Mar 20;56:1339-1344.

[53] Mobaraki S, Zakavi M, Mahmoodi O, Sorkhabadi MO, Khalilinezhad SS, Torkmani RS. An experimental study on the mechanisms of enhancing oil recovery by nanoparticles-assisted surfactant flood. Geosystem Eng. 2020 Nov 1;23(6):315-331.

[54] Mohajeri M, Hemmati M, Shekarabi AS. An experimental study on using a nanosurfactant in an EOR process of heavy oil in a fractured micromodel. J Pet Sci Eng. 2015 Feb 1;126:162-173.

[55] Giraldo J, Benjumea P, Lopera S, Cortés FB, Ruiz MA. Wettability Alteration of Sandstone Cores by Alumina-Based Nanofluids. Energy Fuels. 2013 Jul 18;27(7):3659-3665.

Chapter 4

Petrological and Biostratigraphic Characteristics of Pre-Cenozoic Carbonate Rocks in the Northern Song Hong Basin, Vietnam

Mai Hoang Dam, Nguyen Tan Trieu and Lieu Kim Phuong

Abstract

Pre-Cenozoic carbonate rocks in the northern Song Hong basin, Vietnam that are being considered and studied by oil companies in exploration and exploitation. The hydrocarbon accumulations in these rocks have been discovered and have significantly commercial reserves, in which the porosity plays an important role in estimating the capacity of hydrocarbon. The carbonate rocks are composed mainly of crystalline limestone, packstone, wackestone and mudstone, which have been experienced dolomitization, compaction and dissolution. The main carbonate pore systems include fracture, vuggy and intercrystalline porosity. The predominance of larger benthic foraminiferal assemblages indicates that the carbonate sediments were formed during the late Paleozoic (Carboniferous-Permian) and were deposited in shallow marine environment. Furthermore, the obtained petrological and biostratigraphic characteristics are well-correlated with the carbonate formations exposed in adjacent Cat Ba island area. The results of this study are either used in petroleum exploration or used in a local stratigraphic correlation in northern Vietnam.

Keywords: stratigraphy, lithology, pore types, foraminifera, hydrocarbon, reservoir model

1. Introduction

PetroVietnam recently discovered a hydrocarbon flow in the Pre-Cenozoic carbonate rocks of the northern Song Hong basin, Vietnam northern continental shelf, which has since become a potential object in oil and gas exploration operations and has attracted the interest of petroleum companies. As the Song Hong basin where penetrated by the petroleum exploration wells show that the basement rock formations are mainly sedimentary rocks which were deposited in shallow marine environment. They consist of strongly altered carbonate rocks with age varying from Permian to Early Carboniferous and their thickness reach over 500 m. The above covering of the basement rocks is the Cenozoic sediments and its petroleum systems have been considered within the framework of Cenozoic stratigraphy, this finding has suggested a new approach for managing the petroleum systems of the basin. Therefore, this area has been extensively investigated by PetroVietnam and foreign petroleum companies, and detailed studies have been conducted on the litho-sedimentological characteristics of the carbonate rocks [1–4]; the characteristics of the Mesozoic carbonate reservoir [5]; determining the geological age and building a geological prediction model for the Pre-Cenozoic carbonate basement [6]; and other projects [7, 8] in the northern Song Hong basin. However, geological data for this area is limited, therefore, PetroVietnam has indirectly conducted many studies using field models that focus on the outcrops in areas adjacent to the northern Song Hong basin. One of the problems currently under discussion is the petrological characteristics and ages of the carbonate formations of the carbonate rocks and their relationship in the northern Song Hong basin and the adjacent areas. Many field studies have been conducted to evaluate the petrology, stratigraphy, and tectonic characteristics of the Cat Ba, Co To, and Bach Long Vi islands and their adjacent areas, including Hai Phong, Ha Long Bay, and Quang Ninh. The results of these studies show that the ages of the carbonate formations of the study area in the northern Song Hong basin are similar to those of several islands in the Vietnam northern shelf, according to the characteristics of foraminiferal assemblages. Therefore, this study aims to characterize the petrology and stratigraphy of the carbonate formations in the northern Song Hong basin and their correlation with those of adjacent island. Three wells at a depth of over 3500 m with a carbonate rocks thickness of approximately 500 m were studied in the northern Song Hong basin. This study is very important for providing petrological characteristic and evidence regarding foraminiferal fossils to determine the stratigraphic relationship between wells and adjacent areas. It might also be applied for stratigraphic correlation and comparison of reservoir models in the wells.

2. Geological settings

The study area is located at northern Vietnam on the South china plate (**Figure 1**). In the Late Paleozoic, the South China plate collided with the Indochina plate and formed a broad northwest-southeast mobile belt, which included the Ailaoshan, Song Ma, and Dian-Qiong sutures that represent the complex boundary zone between the Indochina and South China plates in northern Vietnam, and the south-eastern part of the South China plate [9]. The Song Ma suture zone is composed of large amount of serpentinite, altered gabbro, and chromitite. The serpentinite may serve as a remnant of the Paleo-Tethys oceanic lithosphere [10]. According to Metcalfe [11], large-scale folding, thrust, and nappe formation in the Early-Middle Carboniferous, blanketing Middle Carboniferous strata, and plant remains suggest that this suture was originated in the Early Carboniferous.

The northern Song Hong basin is composed of Pre-Cenozoic rocks, including carbonate, clastic, and metamorphic rocks, which are overlain by Cenozoic rocks [12]. Carbonates are the major rocks, while the other rocks are present to a lesser extent. The carbonate rocks consist of about (500 m thick) of Paleozoic successions originating during Devonian and Permian [8, 13–21]. Limestones are classified dolostone and crystalline limestone. The dolostones were probably originally mud-supported limestones that have been partially to completely replaced by variably finely to coarsely crystalline, anhedral to subhedral (xenotopic to hypidotopic), rhombic dolomites and are classified probably as crystalline dolostones. Crystalline limestone, in which lime mud matrix has mostly recrystallized to microspar/pseudospar. Locally, limestone has been fractured. There also appear some dolomite conglomerates, which are predominantly made up of gravel-sized dolomite grains and dolomite cement.

The northern part of this area is in the Quang Ninh zone, the southern part is located in the western part of Bac Bo (Tonkin) Gulf [22], and the entire study area is in the northern Song Ma suture zone. In which, the Quang Ninh zone is studied quite in detail on petrography and paleontology in the sections on Cat Ba island. The exposed lithology on Cat Ba island is dominated by carbonate rocks that have been

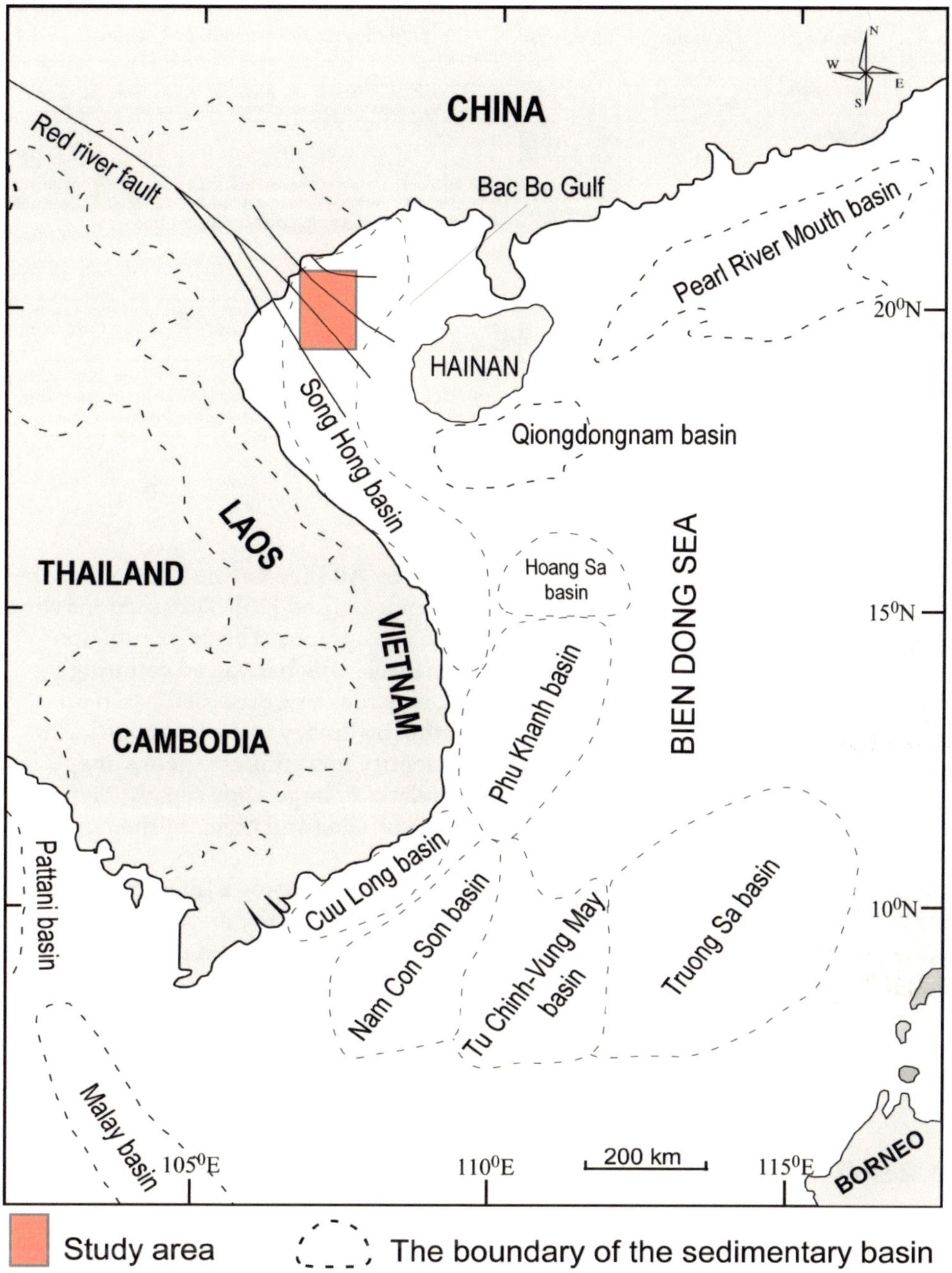

Figure 1.
Map of the sedimentary basins of Vietnam continental shelf and adjacent areas, and the location of study area [4].

described and updated by [13, 20, 23, 24], and consists of the Trang Kenh (D_2-D_3 *tk*), Pho Han (D_3-C_1 *ph*), and Bac Son (C-P *bs*) formations (**Figure 2**).

3. Materials and methods

The current study was performed using petrographic microscopy, scanning electronic microscopy (SEM) and X-ray diffraction (XRD) analysis. 81 thin-sectioned

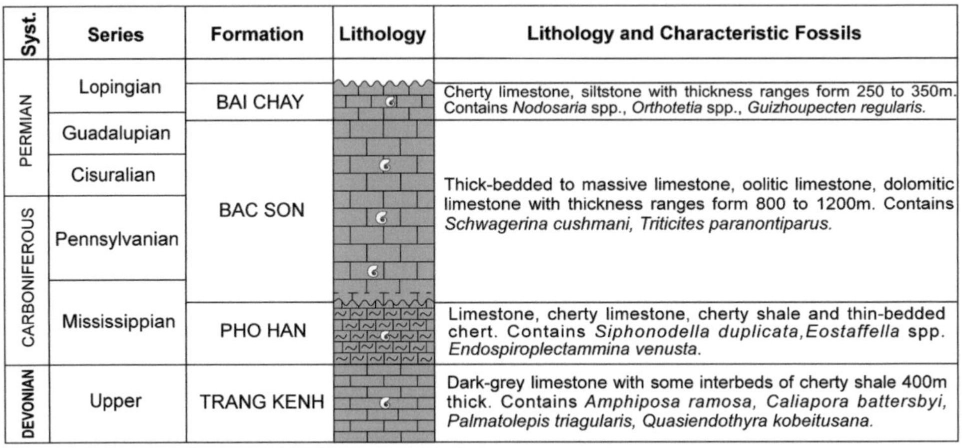

Figure 2.
The late Paleozoic stratigraphy column of the Northeast Vietnam (Bac Bo Gulf) [25].

from carbonate units were prepared and stained by ARS (Alizarine Red Solution) to distinguish calcite and dolomite using Dickson's method [26]. Thin sections were studied under polarized microscopy to analyze petrography. The determination of visible porosity was performed by modal analysis, which involved counting 300 points per thin section [27, 28]. The carbonate rock was classified based on Dunham's classification [29] and its modification by Embry and Klovan [30]. 218 additional oriented thin sections are used to identify foraminifera (genus or species names). If the foraminifera were found to be relatively large upon separation from the carbonate debris, they were fixed onto the glass slide and then polished until all chambers or internal structures could be observed.

31 samples of dolostone and limestone were examined using a JEOL Scanning electronic microscope (SEM) in order to identify and assess the morphology, type of authigenic minerals and their relationship to framework grains and pore network.

93 samples of dolostone and limestone were examined using D8-Advance automatic system that carried out X-ray diffraction (XRD) analysis for determining the mineralogical composition based on amount in term of Semi-Quantitative. All of them were performed at Vietnam Petroleum Institute (VPI) in Vietnam.

4. Results and discussions

4.1 Petrological and stratigraphic characteristics of carbonate rocks in the early Carboniferous

The Early Carboniferous carbonate rocks has sporadically been interbedded dolostone, limestones and dolomitic limestones and a small amount of limestone alternating clamps, with a thickness of approximately 400 m to belong to the Pho Han (D_3-C_1 *ph*) formation. Dolostones consist of euhedral and subhedral rhombic dolomite crystals with planar-euhedral and planar-subhedral texture [31] (**Figure 3**). These rocks have severely been affected by compaction, which are manifested by stylolitization and fracturing. Foraminifera fossil could be not found due to dolomitization. Limestone is classified mainly of lime-mudstone and minor amount of packstone. The lime-mudstones are composed of micrite and microspar calcite. The skeletal grains consist of mainly well-preserved benthic foraminifera and bioclasts.

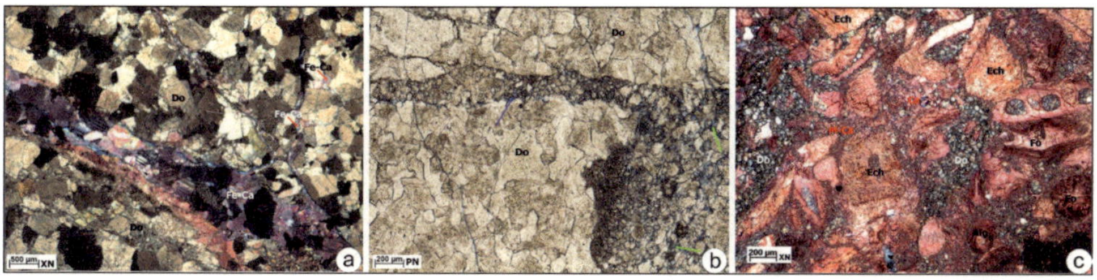

Figure 3.
a. Dolostone is composed of mainly dolomite with euhedral and subhedral rhombus, made up of planar-euhedral and planar-subhedral. Ferroan calcite (Fe-Ca) is remained in dolostone and fill in the fractures; b. Dolostone is composed of mostly subhedral rhombic dolomite (Do). The visible porosity appears in between dolomite crystals; c. Packstone interbeds in dolostone at the 3750 m and 3900 m depth. Packstone is consisted of fossils including echinoderm (Ech), foraminifera (Fo), algae (Alg), bio-fragments and non-ferrous, micrite calcite (stained in pink color). The fossils contacts together and micrite dolomite (Do) appears either inner fossils or on matrix.

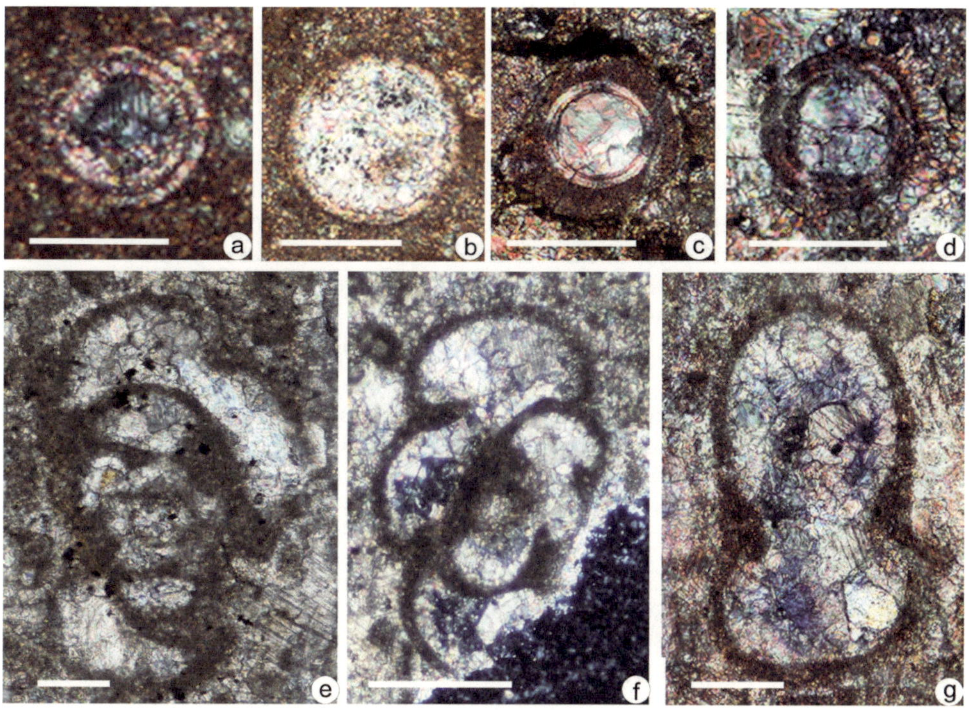

Figure 4.
The characteristic foraminiferal assemblage of the lower middle Tournaisian in the Song Hong basin; a–d. CALCISPHAERIDAE (Williamson, 1880); e. Septabrunsiina kazakhstanica *(Reitlinger, 1961); f.* Septabrunsiina *sp. (Lipina, 1955); g.* Tournayella (?) *sp. (Dain, 1953). Scale bar is 100 μm.*

Two foraminiferal assemblages were identified in the studied interval which is from a depth of 3500–4000 m. The lower part contains unilocular foraminifera, while the upper part consists of Tournayellids. The unilocular foraminifera group characterized by abundant *Calcisphaera* and *Parathurammina*. This assemblage is considered to represent the upper region of the unilocular intermittent zone [21] and indicates that the carbonate formation was originated during the latest Early Tournaisian (**Figure 4a–d**). A fossil assemblage of Tournayellidae, include *Tournayella*, *Septabrunsiina kazakhtanica*, and *Septabrunsiina* sp. This assemblage was characterized by the clear evolution of septa between the chambers, which are affiliated with the lowermost part of the *Chernyshinella-Palaeospiroplectammina* zone in the Cat Co and Gia Luan sections. This suggests that this assemblage was formed during Early to Mid-Tournaisian (**Figure 4e–g**).

Generally, the carbonate of this formation has been strongly replaced by fossil remains that are found scattered in the limestone layers and the visible porosity is noted as retained fracture pores and intercrystalline pores.

4.2 Petrological and stratigraphic characteristics of carbonate rocks in the late Carboniferous

The Late Carboniferous carbonate rocks appear at the depth roughly 3400–4150 m and is consisted of limestones, dolostones and crystalline limestones. At the upper part, the carbonate rocks are interbedded with basalt tuffs and with silic dikes/veins at the lower.

Almost all carbonate rocks are mud-supported type. Limestones are classified mostly as dolomitic-calcitic mudstone, wackestone and packstone. The allochems mainly consist of Foraminifera (fusuline), algae, coral and echinoderm (**Figure 5**), while the groundmass are micrite and microspar. The Lime mud matrix (micrite) has been partly to totally replaced by finely to coarsely crystalline, anhedral to subhedral dolomites (non-ferroan dolomite) and sparry calcite with calcite crystals ranging 10–15 μm diameter (as ferroan calcite, up to 35% and non-ferroan calcite, up to 70%). The carbonates have more or less been replaced by quartz. They have also suffered compaction in the form of fractures; however, all fractures have been filled by sparry calcite, sparry ferroan calcite and silica cement (**Figure 6c** and **d**).

Dolomitic-calcitic limestone that is lime mud matrix recrystallized into microcalcites and replaced by dolomite, was probably originally mud-supported limestone that has been strongly recrystallized to microspar carbonate crystals. The rock consists of abundant carbonate fragments with trace skeletal particles as foraminifera, echinoderm that floating on micrite carbonate matrix. A small amount of very fine to fine sand-sized secondary quartz grains are present. The rock has been undergone the compaction and dissolution in post-deposition. As results, the fractures crossed throughout the carbonate rocks forming stylolite texture and fractures; however, fractures are occluded by calcite and ferroan calcite; minor fracture pores are preserved.

Carbonate allochems consist of algae, foraminifera, and echinoderm. All fossil skeletons have been undergone micritization, however, their cellular structures are locally preserved. Commonly, allochems and micritic carbonate have strongly been recrystallized to with minor amount of ferroan calcite on lime mud matrix.

Figure 5.
Thin section photography of carbonate rock in the Song Hong basin. a, b. both wackestone and packstone consist of foraminifera (Fusuline, Fo) and crinoids (white arrows), which are floating in the lime mud matrix. All crinoids have strongly been undergone calcitization. The lime mud matrix has locally been replaced by – Calcite minerals. Compaction represented by fractures and stylolities, which are filled up by ferroan calcite cements; c. volcanic dikes (Vol) and basalt tuffs (Tuf) cross-cutting both grains and matric.

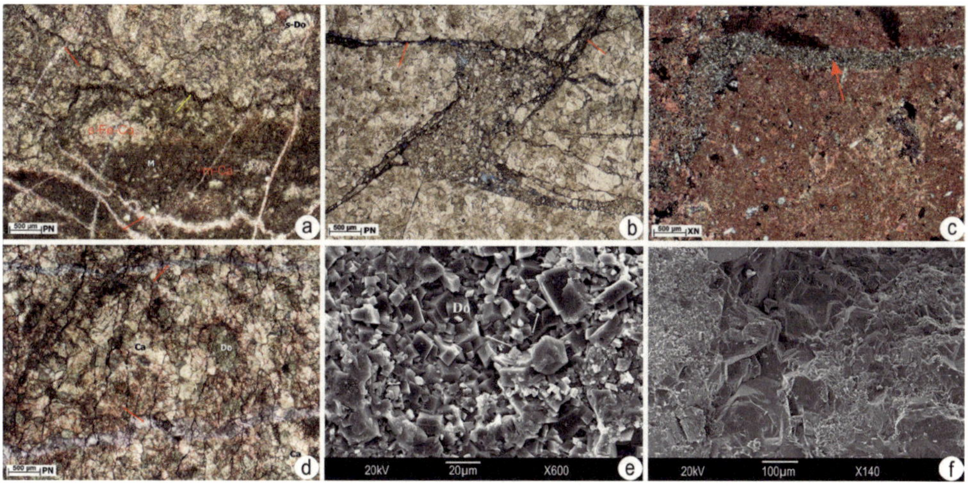

Figure 6.
Thin section and SEM photography of carbonates in the Song Hong basin. a, b. the carbonate rocks have strongly been undergone tectonic activity, resulting in crushed fracturing (red arrows) and complex stylolization (yellow arrow); c. the fractures have been filled up by calcite, chert and ferroan calcite (red arrow); d. the fractures have been filled up chert as seen authigenic quartz and chalcedony then calcite grows (Ca) in remained pores; e. the development of rhombic dolomite crystals, creating many of intercrystalline pores; f. limestone has been completely crystallized into variably coarsely crystalline, anhedral to subhedral calcite and locally replaced with finely rhombic crystallines of dolomite. The sample has been fractured and fractures have been filled up by calcite minerals (along the middle).

The rock has been fractured and almost all fractures have been filled with calcite or ferroan calcite. The size of fractures changes from 0.02–0.4 mm in width.

Crystalline limestones are mainly consist of lime mud that is recrystallized into calcite. Minor allochems are present as echinoderm. The rock have been undergone compaction and dissolution that formed stylolite and fractures.

Dolostone has mud-supported limestone that partially to completely replaced by finely to coarsely crystalline, anhedral to subhedral (xenotopic to hypidotopic), rhombic dolomites. The size of dolomite crystals commonly range from 0.15–0.4 mm, and exhibit equigranular texture. Dolostone has been compacted and formed stylolite and fracture system complication (**Figure 6a** and **b**). It indicates that this rock has been affected by compaction during burial diagenesis. Fractures have been occluded by dolomite crystals.

The result of Scanning Electric Microscope (SEM) analysis show that displays the morphology coarsely crystalline, anhedral to subhedral calcite and locally replaced with finely rhombic crystallines of dolomite and rhombic dolomite crystals, creating many of intercrystalline pores (**Figure 5e** and **f**). The XRD result for whole-rock shows rock-forming minerals of carbonate basement, in which calcite and dolomite are homogeneously present in high levels. Quartz is also sporadically present with high amount. K-feldspar and plagioclase are nearly totally disappeared.

Diagenetic processes, such as micritization, may be contemporaneous with diagenetic process such as cementation. Diagenesis has included micritization of bioclasts and infiltration of micrite calcite into the foraminifera chambers; after that recrystallisation of micrite into microspar and pseudospar sizes; fracturing and non-selective dissolution of calcite, followed by precipitation of blocky calcite cement and locally formed of xenotopic to hypidiotopic dolomite.

In general, the visible porosity of carbonate rocks is mostly formed by dolomitization, dissolution and fracture. It is present as intercrystalline pores, which is formed from the dolomitization; whereas, fracture pores have been created by tectonic activity; vuggy pores that have been formed by the shrinkage of matrix and dissolution of skeletal debris; with minor moldic pores. Carbonate has more or less been fractured; however, almost all fractures have been filled up with silic and calcite due to hydrothermal activity. Therefore, the reservoir quality of the rock has been restricted.

The petrographic analysis result indicated that limestones are classified mostly as crystalline dolomite, dolomitic-calcitic mudstone, wackestone and packstone. These limestones are predominantly made up of carbonate allochems as Foraminifera (as Fusuline), Algae, Coral and Echinoderm, which have been stylolitised and fractured due to tectonic activity. The rocks have also been strongly altered with silic, which resulting from volcanic activity in post-deposition. It is vital for pointing to the depositional settings that were frequently in low to moderate-energy flow and affected by sea-level fluctuation and deposited in a reef-continental shelf shallow marine environment. This reef has more or less been affected by tectonic activity and volcanic activity, in which carbonate has been fractured, stylolitised and altered with silic minerals.

Two foraminiferal assemblages were identified in the studied interval which characterizes the Serpukhovian-Early Bashkirian and Late Moscovian age. The *Millerella-Eostaffella* zone was observed in the ~150 m thick limestone formations of the lower section of the basement. This zone was characterized by the abundant Pseudostaffellinae including: *Eostaffella, Neostaffella, Mediocris,* and *Pseudoendothyra,* common *Palaeotextularia,* and sudden disappearance of Endothyrinae at the

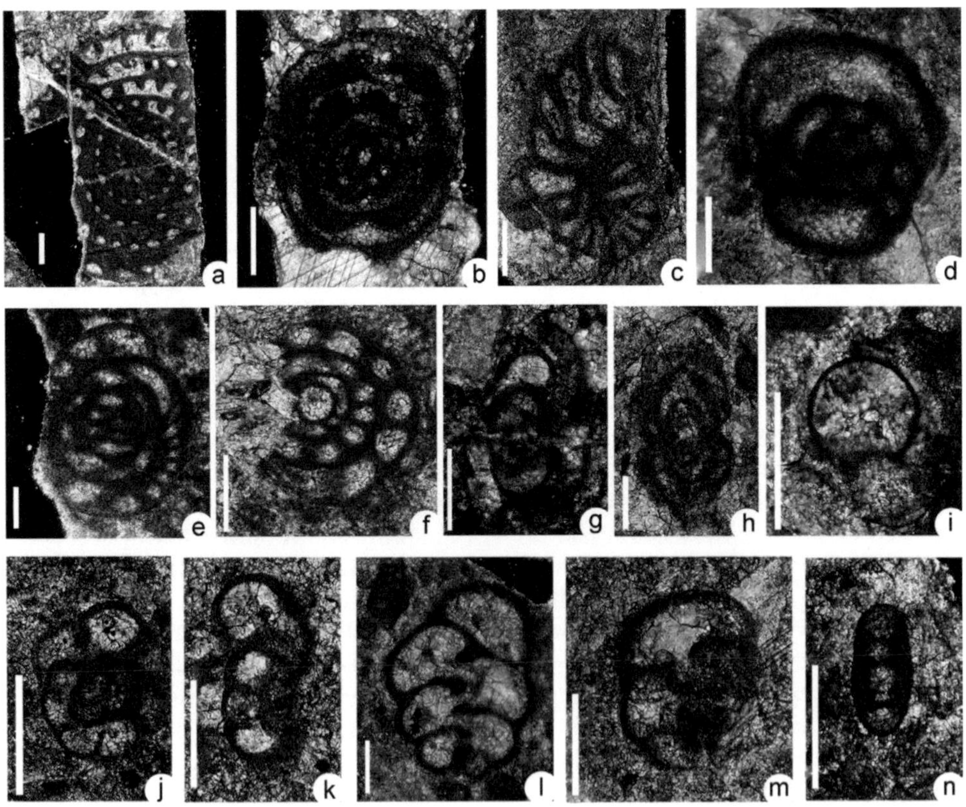

Figure 7.
The characteristic foraminiferal assemblage of the Serpukhovian-Moscovian in the Song Hong basin. a. Profusulinella *sp. (Rauzer-Chernousova and Belyaev, 1936); b.* Neostaffella *sp. (Miklukho-Maklay, 1959); c. Fusulinid fragment; d.* Grovesella *sp. (Morelet, 1969); e.* Schubertella *sp. (staff and Wedekind, 1910); f.* Fusulinella *sp. (Möller, 1877); g, h.* Pseudoendothyra *spp. (Mikhaylov, 1939); i.* Eotuberitina *sp. (Miklukho-Maklay, 1965); j, k. ENDOTHYRINAE (Brady, 1884); l.* Palaeotextularia *sp. (Simakov, 1992); m.* Globivalvulina *sp. (Simakov, 1992);* n. Mediocris *sp. (Rozovskaya, 1961). Scale bar is 100 μm.*

end of the Serpukhovian [32]. The upper region of this zone contained the first *Profusulinella* and Ozawainellidae, which are indicative of the upper Bashkirian [14, 21] (**Figure 7**). The foraminiferal assemblage of this zone is similar to that in the uppermost Serpukhovian to upper Bashkirian in the Gia Luan section.

The *Fusulinella-Fusulina* zone was found in the ~100 m thick limestone in the middle section of the basement. It was identified by the abundances of *Profusulinella, Fusulina, Schubertella*, Ozawainellidae, *Globivalvulina*, and *Eotuberitina* (**Figure 7**). This zone is indicative of the upper Moscovian. In addition, calcareous algae groups are abundant in this zone, with *Beresella* appearing as the most predominant genus. This genus first appeared at the end of the early Carboniferous (Serpukhovian) and was the most abundant in the first half of the late Carboniferous, from the Bashkirian to the Moscovian [33, 34].

4.3 Petrographical and stratigraphic characteristics of carbonate rock in the middle-late Permian

In this studied section, the Early Carboniferous carbonate rocks appear at the depth roughly 3505–4050 m and is consisted of wackestone, packstone and mudstone to belong to Bac Son (C-P *bs*) formation. At the lower part, the carbonate rock is verified mainly as packstone with grain supported and contains many foraminiferal fossils. The majority composition minerals of limestone is non-ferroan calcite (stained in pink), locally ferroan calcite (stained in mauve), and ferroan dolomite. The limestones contain carbonate allochems as benthic foraminifera, echinoderm, coral, algae, bryozoa, brachiopod. The limestones are highly fractured and stylolitised owing to tectonic activity, squeezing, diagenesis, that filled up by ferroan calcite, dolomite (**Figure 8**).

After silicate (chalcedony) depositional process, the crystalline minerals filled up the fractures which is a dyke intruding into the limestone and nodules fill in the fractures. The visible porosity of carbonate rock is created by the dissolution of vuggy pores and fractured pores that locally preserved. Carbonate basement rock is classified as after Dunham's classification [29], analyzed limestones are verified as wackestone, packstone and mudstone types.

Packstone contains a fair level of fossils such as foraminifera, algae, echinoderm, coral, bryozoa, brachiopod and other bio-fragments that contains more than 10% in total rock composition. Carbonate allochems and bio-fragments contact together and cemented by lime mud that is crystallized to microspar calcite (4–10 μm), pseudospar calcite (10–50 μm) and locally dolomitised. All fossil skeletons have been completely altered by calcite, their cellular structures are also altered by calcite

Figure 8.
Thin section photography of Permian carbonate in the Song Hong basin. The main components are non-ferroan micrite calcite (Mi-Ca, stained in pink) intermixing with minor amount of organic matter (Org), locally replaced by dolomite (white arrows) and quartz. The rock contains bio-fragments such as foraminifera (Fo), ostracod (Os) and unidentified bio-fragments (Bio). The rock has been squeezed and created fractures (blue arrows), filled up by calcite (Ca).

and locally well preserved. Limestone has been fractured but the fracture pores are occluded by calcite (stained in pink) and ferroan calcite (stained in mauve).

Wackestone contains smaller amount of fossils such as foraminifera, ostracods, algae other bio-fragments that contains about 10% in total rock composition. Bio-fragments are floating on lime mud matrix that micrite texture and locally altered into dolomite and replaced by silicite. All fossil skeletons and their cellular structure are altered and filled up by calcite. Limestone has been fractured, stylolite texture, filled up by non-ferroan calcite.

Mudstone contains dominantly lime mud, micrite texture that is crystallized into microspar calcite and micrite calcite (<4 μm). Carbonate allochems as foraminifera, ostracods and unidentified bio-fragments that contain less than 10% in total rock composition. Carbonate allochems are floating on lime mud matrix, and locally altered into dolomite and replaced by silicate.

Dolostone has been formed from packstone, wackestone that dolomitised in the alteration post-depositional process and interbedded in packstone and wackestone. Dolostone have planar-subhedral texture and carbonate allochems in dolostone have been dolomitised.

The result of (SEM) analysis reveals the crystalline morphology of micrite calcite (Ca), size <4 μm and rhombic dolomite (Do) with euhedral with size >50 μm (**Figure 9e** and **f**). Together with the whole rock XRD results indicated that the most predominant volume is carbonate minerals in which mainly calcite and less than as dolomite, rarely siderite. Minor amount of quartz, feldspar minerals are also found at this interval.

The petrographic analysis result in this study shows limestone experienced the alteration post depositional process such as the crystalline of lime mud altered into calcite, dolomitised. The squeezing and dissolution process created fractures and stylolite textures, locally fractures filled up by calcite, dolomite and silicate.

Lime mud has been crystallized and altered into calcite; Whereas, lime mud in bio-fragments crystalline altered into micrite calcite and sparry calcite that surrounded bio-fragments and created poikilotopic texture.

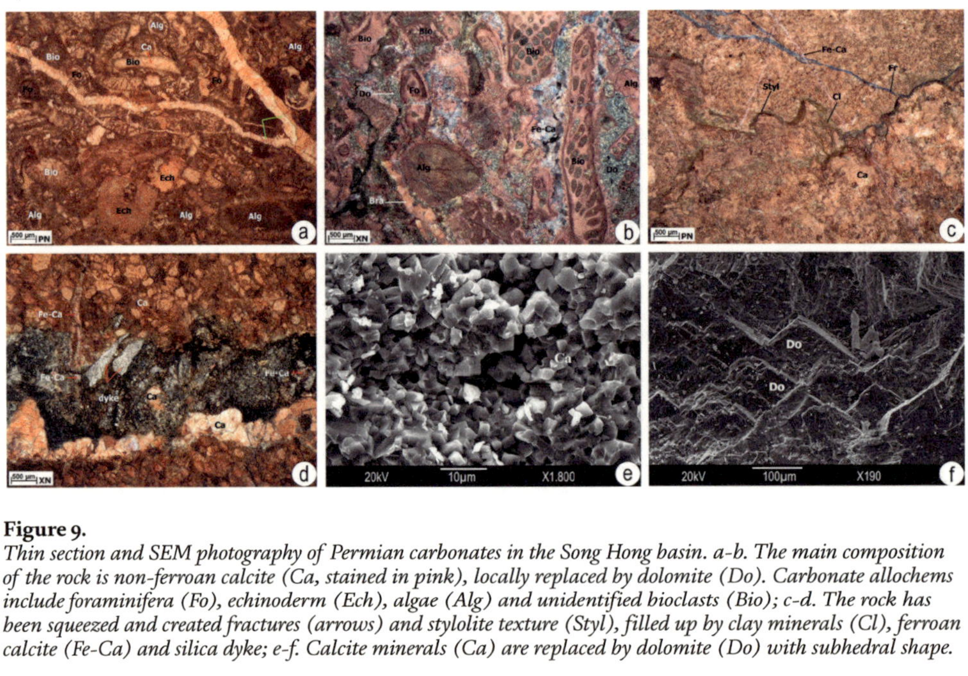

Figure 9.
Thin section and SEM photography of Permian carbonates in the Song Hong basin. a-b. The main composition of the rock is non-ferroan calcite (Ca, stained in pink), locally replaced by dolomite (Do). Carbonate allochems include foraminifera (Fo), echinoderm (Ech), algae (Alg) and unidentified bioclasts (Bio); c-d. The rock has been squeezed and created fractures (arrows) and stylolite texture (Styl), filled up by clay minerals (Cl), ferroan calcite (Fe-Ca) and silica dyke; e-f. Calcite minerals (Ca) are replaced by dolomite (Do) with subhedral shape.

Dolomitised shows that micrite calcite has partly been altered by dolomite and locally filled up fractures. Dolomite crystals are more anhedral formed after that, in buried stage.

Fracture and stylolite have been formed by tectonic activity and they were filled up by calcite, dolomite.

Generally, the visible porosity is mainly fractured pores, vuggy and moldic pores intergranular pores between dolomite crystals. Because of the impaction hydrothermal activity, vuggy, fracture that infilled by silica dyke and nodule into limestone and restricted the reservoir quality of the rock, so the fracture pores are estimated in poor.

Based on the petrographic analysis result, the presence of fossils such as especially foraminifera group in mudstone and wackestone that contain lime mud, some bio-fragments and dominantly deposited in low energy marine, steady current. Packstone contains abundant carbonate allochems that have been deposited in marine environment with the changes of energy current from low to high and on the contrary.

The fossil skeletons were mainly replaced by calcite cement and dolomite and the inside of the chambers were also dolomitised but the structure of the skeleton is still very well preserved. The fossil assemblages found to characterize Late Paleozoic (Permian) is mainly distributed in northeast Vietnam and adjacent areas [21, 35]. The typical representatives include genera: *Nodosinelloides, Nodosaria, Geinitzina, Codonofusiella, Pachyphloia, Rectoglandulina, Palaeotextularia, Reichelina, Cribrogenerina* and popularity of fossil fragments of the Fusulinacae superfamily were dominant in samples, which exhibit strong folding septa and very clear keriotheca that are typical of the Schwagerinidae family (**Figure 10**), and are mainly distributed from the Kasimovian (Late Carboniferous) to the Capitanian (Middle Permian) [36–38]. In addition, the presence of the *Geinitzina, Pachyphloia, Nodosinelloides*, and *Cribrogenerina* genera in this well was also characteristic of the Permian [39, 40]. Moreover, algal groups mainly consisting of Konickopora were abundant. The first representative Konickopora in southern China were observed in formations from the Serpukhovian, and they became abundant in the Permian. The abovementioned foraminiferal assemblages suggest that this carbonate formation originated during the Capitanian-Wuchiapingian.

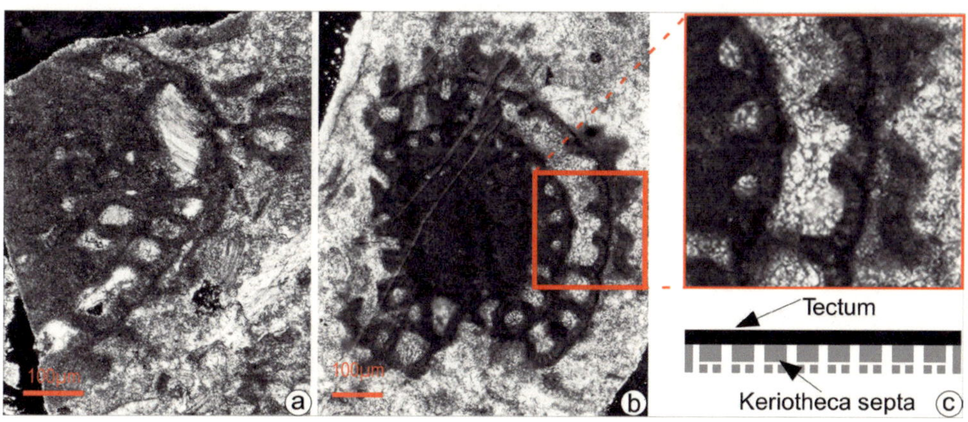

Figure 10.
a, b. Characteristics of the keriotheca septa of the Schwagerinidae family found in the study well; c. detailed structure of the keriotheca septa.

According to previous studies [21, 35] which were recorded the distribution of foraminifera assemblages in many areas on the Vietnamese continent and adjacent areas (**Figure 11**).

Genus *Globivalvulina* Schubert, 1921 was distributed stratigraphy from Serpukhovian to the latest Permian [41]. Genus *Reichelina* Erk, 1942 was identified in the range from Wuchiapingian to Changhsingian in South China [41]. In northern Vietnam, the distribution stratigraphy of *Reichelina* in Late Permian was found mainly in the Northeast to belong to Bac Son (Wuchiapingian) and Dong Dang (Changhsingian) Formation [21].

Genus *Pachyphloia* Lange, 1925 appeared first from Sakmarian and disappeared in the latest part of Permian in South China [41]. In Vietnam, *Pachyphloia* has been

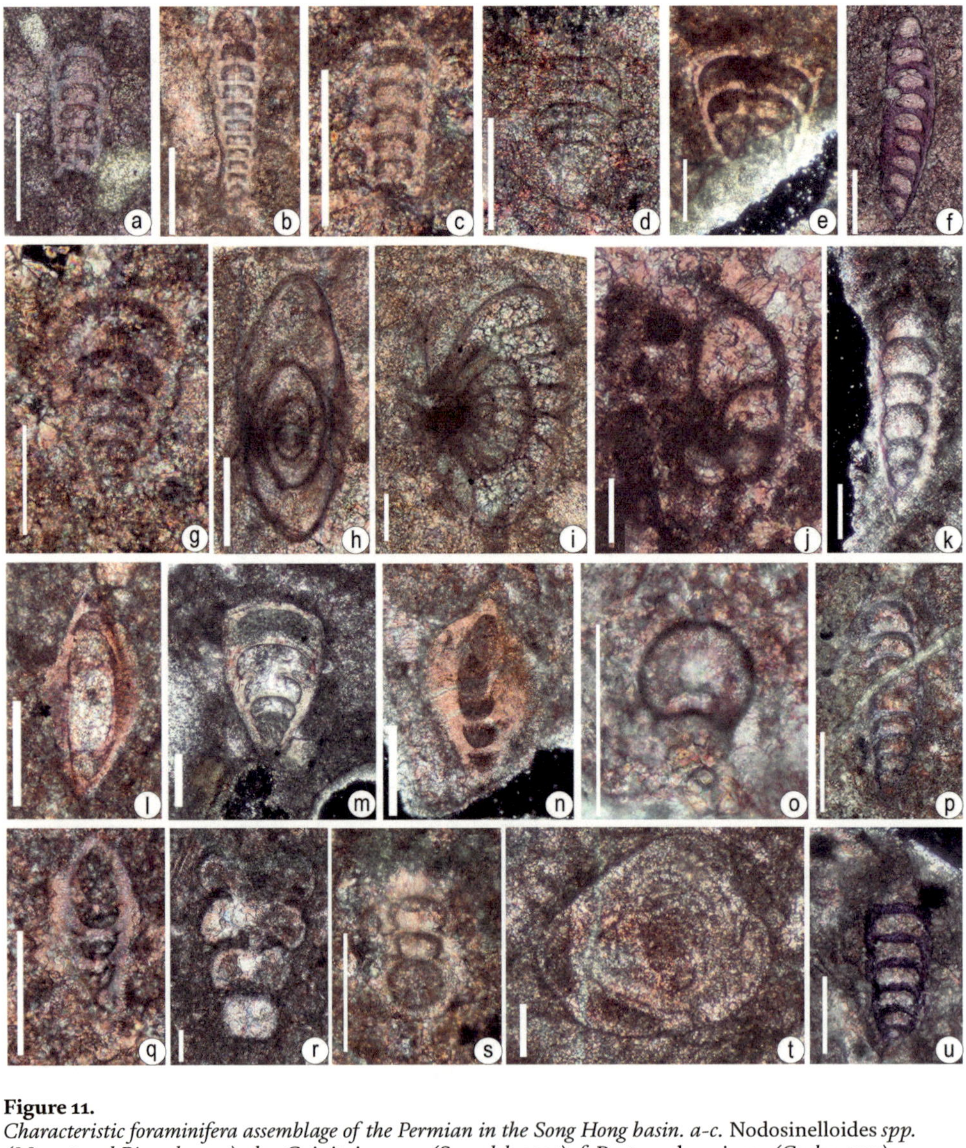

Figure 11.
Characteristic foraminifera assemblage of the Permian in the Song Hong basin. a-c. Nodosinelloides *spp. (Mamet and Pinard, 1992), d-e.* Geinitzina *spp. (Spandel, 1901), f.* Protonodosaria *sp. (Gerke, 1959), g.* Geinitzina *sp. (Spandel, 1901), h-i.* Sichotenella *spp., j.* Globivalvulina *sp. (Schubert, 1921), k.* Protonodosaria *sp. (Gerke, 1959), l.* Pachyphloia *sp. (Lange, 1925), m.* Pachyphloia *sp. (Lange, 1925), n.* Pachyphloia *sp. (Lange, 1925), o.* Diplosphaerina *sp. (Derville, 1952), p.* Protonodosaria *sp. (Gerke, 1959), q.* Pachyphloia *sp. (Lange, 1925), r.* Cribrogenerina *sp. (Schubert, 1908), s.* Nodosaria *sp. (Lamarck, 1812), t.* Neodiscus *(?) sp. (Miklukho-Maklay, 1953), u.* Protonodosaria *sp. (Gerke, 1959). Scale bar is 100μm.*

commonly found from the Middle to Late Permian and widespread in the northeast area to belong to the Bac Son and Dong Dang Formations [21].

Genus *Schubertella* Staff and Wedekind, 1910 was commonly found in the northern Vietnam (Bac Kan, Quang Ninh, Thai Nguyen, Quang Binh) in the range from Late Carboniferous (Moscovian) to Permian to belonging to the Bac Son Formation [21]. Genus *Cribrogenerina* Schubert, 1908, distributed in Late Permian (Changhsingian) of the Dong Dang Formation, was found in Cao Bang, Lang Son [21]. Family Schwagerinidae Dunbar et Henbest, 1930 is characteristic to the Permian of the Bac Son Formation in the North and the South of the Ha Tien Formation [21].

An assemblage of *Nodosinelloides–Geinitzina* which was characteristic of the Early Permian was recorded in Iran [42]. *Codonofusiella–Reichelina* assemblage is abundant in the northeast to belonging to the Bac Son Formation (Wuchiapingian). Genus *Codonofusiella* Dunbar et Skinner, 1937 distributed in the Middle-Late Permian to belong to the Bac Son, Dong Dang, Ha Tien Formations.

5. Conclusions

The carbonate rocks in the study area were originated during Late Paleozoic and have been determined by the foraminiferal fossils which show stratigraphic distribution ranges from the Early Carboniferous (Tournaisian age) to the Late Permian (Capitanian-Changhsingian age). These assemblages reveal the stratigraphic relationship that exists between the carbonate formations in Cat Ba island and the basement rock in the northern Song Hong basin, and provides chronostratigraphic data that can be used in geological models used for hydrocarbon exploration. Most of the carbonate rocks in the Carboniferous period are major dolostone and minor crystalline limestone, wackstone and packsonte.

The carbonate rocks have undergone the post-depositional alteration. The rock fabric and rock composition have changed as lime mud changes micro and sparry calcite with larger size and calcite replaced dolomite which is different component. Dissolution of minerals in the chemical diagenesis leave pores. Additionally, calcite micro and sparry are replaced by silic materials as quartz. The carbonate rocks are highly fractured and compacted forming fractures and stylolites; however, the fractures are infilled by calcite and dolomite in the later diagenesis and the fractures are also blocked by silica materials owing to hydrothermal activity.

Acknowledgements

The authors gratefully acknowledge the Vietnam Petroleum Institute (VPI) for permission to use their data and samples for this paper. The authors would like to thank the Petrology and Biostratigraphy Labs. for supporting valuable analysis to complete this study.

Conflict of interest

All authors have participated in (1) conception and design, or analysis and interpretation of the data; (2) drafting the article or revising it critically for important intellectual content; and (3) approval of the final version.

The authors have no affiliation with any organization with a direct or indirect financial interest in the subject matter discussed in the manuscript.

Author details

Mai Hoang Dam[1*], Nguyen Tan Trieu[1] and Lieu Kim Phuong[2]

1 Vietnam Petroleum Institute, Hanoi, Vietnam

2 Ho Chi Minh City Institute of Resources Geography - Vietnam Academy of Science and Technology, Ho Chi Minh, Vietnam

*Address all correspondence to: dammh@vpi.pvn.vn

References

[1] Tam LT, Hoang CM, Tuan PV. Litho-sedimentological characteristics of Pre-Cenozoic carbonate in Ham Rong field, northeast of Song Hong basin. PetroVietnam Journal. 2014; 5:23-30.

[2] Phuong LK, Luan BT, Tuyen VT. Petrographic studies and diagenetic evolution of Paleozoic carbonate basement rock in the northwest Song Hong basin. Vietnam Journal of Science, Technology and Engineering. 2019; 61(8):1-6

[3] Dam NH, Tuyen VT. Characteristic of petrology and distribution of larger benthic foraminifera of Permian carbonate in the southern part of block 106, Song Hong basin. PetroVietnam Journal. 2020; 3:22-31

[4] Dam MH, Phuong LK, Vo NVS, Tham NT, Huy VV. Characteristic foraminifera of the Pre-Cenozoic carbonate formations of Cat Ba island and northern Song Hong Basin, Vietnam. Marine and Petroleum Geology. 2020; 120:1-24. DOI: 10.1016/j.marpetgeo.2020.104543

[5] Tam LT, Tuan PV, Hung NV. Characteristics of Mesozoic carbonate reservoir in Ham Rong field, block 106 in Vietnam's continental shelf. PetroVietnam Journal. 2015; 5:26-31

[6] Huy TN, Trung ND, Ninh TD, Thang NV. Determination of geological age and predict of geological model for Pre-Tertiary carbonate basement in the north Bac Bo Gulf, Vietnam. PetroVietnam Journal. 2014; 11:29-40

[7] Hoang NV, Duc NA. A study of fractured carbonate in Blocks 102-106 by wireline logging. PetroVietnam Journal. 2014; 11:23-28

[8] Vo NVS, Dam MH. Atlas of foraminifera of the pre-Cenozoic carbonate in Cat Ba island and wells in northern Song Hong basin [project]. Hanoi: Vietnam Petroleum Institute; 2016.

[9] Metcalfe I. Palaeozoic-Mesozoic history of SE Asia. Geological Society. 2011; 355(1):7-35. DOI: 10.1144/SP355.2

[10] Trung NM, Tsujimori T, Itaya T. Honvang serpentinite body of the Song Ma fault zone, Northern Vietnam: A remnant of oceanic lithosphere within the Indochina–South China suture. Gondwana Research. 2006; 9:225-230.

[11] Metcalfe I. Palaeozoic and Mesozoic geological evolution of the SE Asian region: multidisciplinary constraints and implications for biogeography. In: Hall R, Holloway JD, editors. Biogeography and Geological Evolution of SE Asia. Amsterdam: Backhuys Publishers; 1998. p. 25-41.

[12] Huyen NM, Hoai DH. Song Hong basin and petroleum resources. In: Hiep N, editor. Geology and Petroleum resource in Vietnam. Hanoi: Science and Technics Publishing House; 2007. p. 186-239.

[13] Liem NV. The Carboniferous in the southern Vietnam. Journal of Biology–Geography XVI. 1978; 3:78-85.

[14] Liem NV. Upper Paleozoic in Vietnam. Hanoi: Sciences and Technics Publishing House; 1985. 79 p.

[15] Thanh TD. The Devonian in Vietnam. Hanoi: Science and Technics Publishing House; 1986. 144 p.

[16] Thanh TD, Khuc V. Stratigraphic units of Vietnam. Hanoi: Vietnam National University Press; 2005. 504 p.

[17] Phuong TH, Truong DN. Initial results of Devonian-Carboniferous boundary study in the southern Cat Ba section, Hai Phong. Vietnam National

University in Journal of Science XXI. 2005; 4:38-47.

[18] Phuong TH, Truong DN. Discussion about Devonian-Carboniferous boundary in the southern Cat Ba island. Journal of Geology. 2007; 298:12-17.

[19] Phuong HT, Hoa TT, Thanh TD, Cu NH. Geodiversity in the Cat Ba island-A basis for establishing a geopark. Vietnam Journal of Earth Sciences. 2009; 31(3):236-247.

[20] Truong DN. On the stratigraphic stratification of Late Devonian-Early Carboniferous sediments in the coastal region of the northeastern Vietnam. Journal of Geology. 2003; 27:1-9.

[21] Truong DN. Atlas of paleontology in Vietnam (the part of foraminifera). Hanoi: Vietnam Publishing House of Natural Resources, Environment and Cartography; 2012. 119 p.

[22] Tri TV, Khuc V. Geology and Mineral Resources of Viet Nam. Hanoi: Natural Scientific and Technological Publishing House; 2009. 589 p.

[23] Hap NQ. The sediments of the northeastern margin of Hanoi trough and prediction their development into the trough. Journal of Geology. 1967; 69-70:9-21.

[24] Truong DN, Khoa ND. Biostratigraphy of the carbonate sediments in the Upper Devonian-Lower Carboniferous in Quang Binh and Hai Phong areas. Journal of Geology. 1994; 220:1-113.

[25] *Huy TN, Trung ND, Ninh TD, Thang NV. Determination of geological age and predict of geological model for Pre-Tertiary carbonate basement in the north Bac Bo Gulf, Vietnam.* PetroVietnam Journal. *2014; 11: 29-40.*

[26] Dickson JAD. A modified staining technique for carbonates in thin section. London: Nature; 1965. 587 p. DOI: 10.1038/205587a0

[27] Van der Plas L, Tobi AC. A chart for judging the reliability of point counting results. Am J Sci. 1965; 263:87-90. DOI: 10.2475/ajs.263.1.87

[28] Solomon M, Green R. A chart for designing modal analysis by point counting. Geol Rundsch. 1966; 55:844-848. DOI: 10.1007/bf02029658

[29] Dunham RJ. Classification of Carbonate Rocks according to Depositional Textures. American Association of Petroleum Geologist (AAPG) Memoir. 1962; 1:108-121.

[30] Embry AF, Klovan JE. A Late Devonian Reef Tract on Northeasterm Banks Island. Canadian Petroleum Geology. 1971; 19:730-781.

[31] Sibley DF, Gregg JM. Classification of Dolomite Rock Textures. Journal of Sedimentary Research. 1987; 57:967-975.

[32] BouDagher-Fadel MK. Evolution and Geological Significance of Larger Benthic Foraminifera. Amsterdam: Developments in Palaeontology and Stratigraphy 21; 2008. 544 p.

[33] Mamet B, Zhu Z. Carboniferous and Permian algal microflora, Tarim basin (China). Geologica belgica 8. 2005; (1-2):3-13.

[34] Riding R. Calcareous algae and Stromatolites. Berlin: Springer–Verlag; 1991. 571 p. DOI: 10.1007/978-3-642-52335-9

[35] Huyen DT. Stratigraphy of phanerozoic sediments in northeastern Vietnam [project]. Hanoi: Vietnam Institute of Geosciences and Mineral Resources; 2007.

[36] Groves JR, Rettori R, Altiner D. Wall structures in selected Paleozoic lagenide

Foraminifera. Journal of Paleontology 78. 2004; 2:245-256. DOI: 10.1666/0022-3360(2004)078<0245: wsispl>2.0.co;2

[37] Groves JR. Fusulinid wall structure in the *Profusulinella-Fusulinella* evolutionary transition. Bulletins of American Paleontology. 2005; 369:199-218.

[38] Vachard D. New SEM observations of Keriothecal walls: Implications for the evolution of Fusulinida. Journal of Foraminiferal Research 34. 2004; 3:232-242. DOI: 10.2113/34.3.232

[39] Groves JR. Suborder Lagenina and other smaller foraminifers form uppermost Pennsylvanian–Lower Permian rocks of Kansas and Oklahoma. Micropaleontology 46. 2000; 4:285-326.

[40] Loeblich AR, Tappan H. Foraminifera genera and their classification. New York: Van Nostrand Reinhold; 1987. 970 p.

[41] Gaillot J, Vachart D, Galfetti T, Martini R. New latest Permian foraminifers from Laren (Guangxi Province, South China): Palaeobiogeographic implications. Geobios. 2009; 42:141-168.

[42] Yarahmadzahi H, Vachard D, Dibadin B. Smaller foraminifers from the lower permian emarat formation, East of Firuzkuh (Central Alborz, Iran). Research in Paleontology and Stratigraphy. *2016; 122(3):103-118.*

Chapter 5

Petro-Mineralogical and Geochemical Study of the Acid Magmatic Rocks of Tusham Ring Complex, NW Peninsular India

Naveen Kumar and Naresh Kumar

Abstract

The present contribution reports about the field and petrographical observations which are very important to explain the magmatic evolution and geodynamic setting of Tusham Ring Complex (TRC). TRC is associated with A-type acid volcano-plutonic rock-association which is very common characteristics of Neoproterozoic Malani Igneous Suite (MIS). Based on the geological field information, the investigated rock-types are classified as volcanic phase, plutonic phase and dyke phase. Petrographically, rhyolites show porphyritic, granophyric, glomeroporphyritic, aphyritic, spherulitic and perlitic textures whereas granites show hypidomorphic, granophyric and microgranophyric textures. Based on mineral chemistry and whole-rock geochemistry, the petro-mineralogical results are justified and proposed that the rocks under study belong to A-type affinity, within-plate and anorogenic magmatism. Physiochemical features i.e. F and Cl-rich biotite, pegmatite rim, high mineralized veins, micro-granular enclaves and altered mineralogy indicate rock-fluid interactions which are caused by magmatic origin or secondary metasomatic alteration superimposed on the host rock.

Keywords: Tusham Ring Complex, Malani Igneous Suite, A-type, geodynamics

1. Introduction

Tusham Ring Complex (TRC) has been divided into 8 isolated hills i.e. Khanak, Dadam, Tusham, Dharan, Dulheri, Riwasa, Nigana and Devsar [1–4]. All these hills represent sub-volcanic, independent, isolated, elliptical, circular geological settings which display the distinct ring structures which are very common in Malani Igneous Suite [2, 5]. Riwasa and Tusham consist of rhyolite as volcanic phase whereas Khanak, Dadam, Dharan, Dulheri, Nigana and Devsar consist of granite as plutonic phase. The mirco-granular granites and rhyolites are also identified as dyke phase which was intruded in the last phase of magmatism. In the present study, we will discuss only the granitoids of Riwasa, Nigana, Dharan and Dulheri with their field photographs and microscopic results. Being the most abundant rocks in the Earth's upper continental crust, granitoids are extensively studied because they are closely related to with magmatic processes, crustal evolution, tectonics and geodynamics [6, 7]. A-type magmatic suites were recorded from different locations of the world

IntechOpen

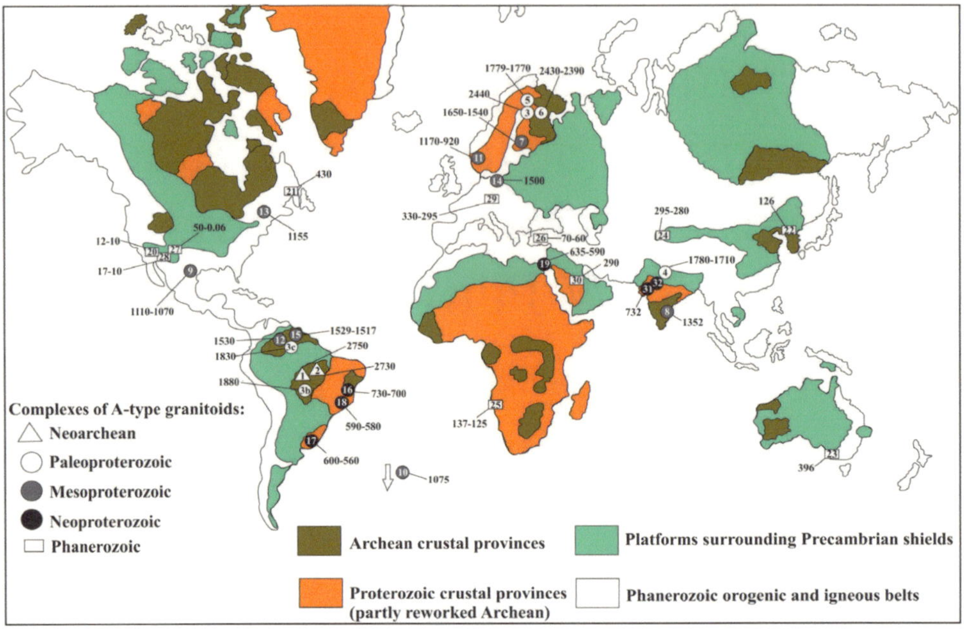

Figure 1.
Global map showing location and complexes of A-type granitoids formed in lithospheric context and relation to crustal evolution. The location number 32 represent A-type suite of Tusham ring complex in NW Indian shield (modified after Haapala and Ramo [8]).

and they are sketched with crustal provinces, platforms, shield areas and orogenic belts with different ages (**Figure 1**). The MIS, NW peninsular India is characterized by isolated, discontinuous, ring-shaped and elliptical outcrops of acid volcano-plutonic rocks with minor outcrops of basic rocks as continental manifestation. The main exposures exit around Siwana, Jalor, Jhunjhunu and Nakora had been extensively explored [9, 10], whereas the MIS exposed in other areas has not been studied in detail. Nevertheless, limited information is available in the literature related to magmatic rocks occurrences in Tusham Ring Complex ([1]; Sharma and Kumar; [2, 4]), so that the purpose of this paper is to provide new field observations and petro-mineralogical data of study areas with respect to MIS.

2. Geological overviews

MIS (bimodal, anorogenic, plume-related, 55,000 km^2 area, 3–7 km thick, ~780–750 Ma) exposed in NW India, is a Precambrian silicic large igneous province, represented by Pan-African thermo-tectonic event [2, 3, 11]. This event indicated multiphase volcanic and plutonic igneous assemblages which were operated by hot spot tectonism during the Neoproterozoic time. A-type magmatic suites are dominant in TAB of NW India, in which felsic rocks are common with alkaline, peralkaline, metaluminous and peraluminous geochemical characteristics [12]. The geological conditions required to erupt such voluminous felsic magma suggest a high rate of magma generation, migration and accumulation in northwestern peninsular India. They are well exposed in Tusham (Haryana), Jhunjhunu, Siwana, Jalor, Nakora, Jodhpur, Mokalsar, Sirohi (Rajasthan) and also in Nagar Parkar (Sind-Pakistan), Kirana (Lahore-Pakistan) areas [1–3, 9]. TRC is peraluminous, within-plate setting and co-magmatic volcano-plutonic granitoids [12].

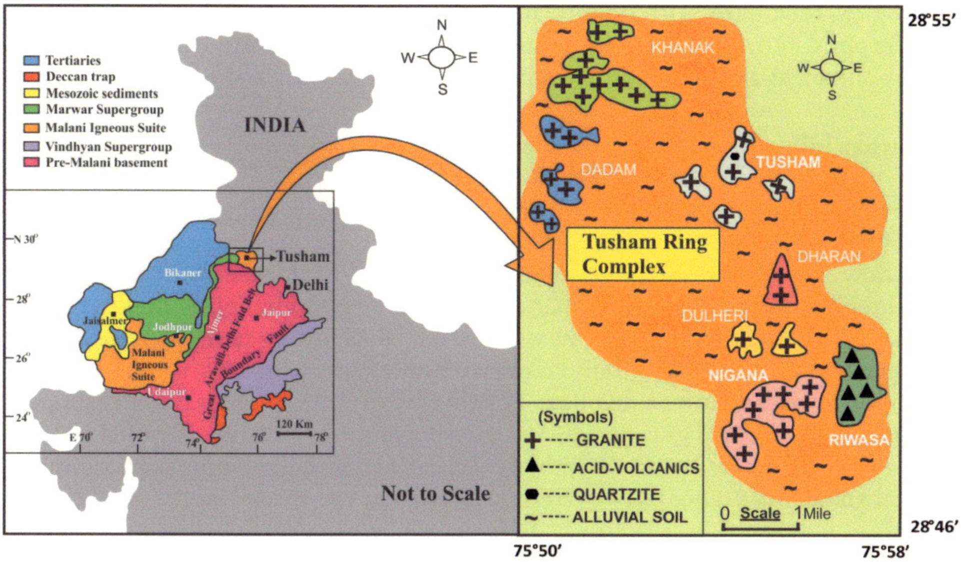

Figure 2.
Sketched map showing the location of the Malani igneous suite in NW India and simplified geological map of the Tusham ring complex in southwestern Haryana, India.

It represents MIS extension in Haryana state of Indian Shield [2], surrounded by independent isolated elliptical hill-locks of granitic and rhyolitic magmas which display the distinct ring structures [2, 5]. These granitoids around TRC are massive and homogeneous with complex geological structures viz.,- xenoliths, post-consolidation joints, fractures, spheroidal weathering and high mineralized veins indicating that they were emplaced in an extensional environment. The present study areas in TRC are located about 160 km WNW of Delhi and far away 400 km NE of Jodhpur (Survey of India topographic sheet no. H43V13; Scale 1: 50,000; 28°47′- 28°49’ N, 75°55′-75°58′ E) (**Figure 2**). Malani rocks in the Tusham area are sandwiched between Delhi quartzite and Vindhyan arenaceous sediments [2, 5]. Various rock-types from different locations are extensively studied to get age (~732 ± 50) of MIS using many isotopic proxies [11, 13–16]. The Malani plume was responsible for the separation of Trans-Aravalli Block (TAB) from East Gondwana, that's why the emplacement of alkali granite and associated acid volcanics having a peraluminous-peralkaline composition in Trans-Aravalli Block are the continental manifestations of plume activity and extensional tectonic regime at 732 Ma [5]. Being a small portion of NW continental block, the field and the petro-mineralogical study of TRC are very important factor to describe the petro-genetic history and geodynamic evolution of MIS.

3. Field observations

Gravity, magnetic and radiometric studies supported triple gravity junction, magnetic anomaly and peak values of HHP around TRC [5]. Gravity and heat flow data are indicative of extensional tectonic environment in the studied MIS region. Paleo-magnetic data also supported the existence of Malani supercontinent which was formed by intraplate, anorogenic, A-type and extensional environment [5, 12]. The seismic, thermal and chemical anomalies in the TAB of NW peninsular India shield is signaling of plume activity in the region. Various lithological rock-suits with field relationship are sketched in field photographs and the petro-mineralogical study that is carried out sincerely. The detailed physio-chemical characteristics of different hills are described as:

3.1 Riwasa hill

About 1200 meter long and 600 meter wide NE–SW trending rhyolites are exposed at Riwasa. It has mainly gray and pink color and shows apparently magmatic flow. Rhyolitic dykes are of varied dimensions (0.5–4 meter) cut across the gray and pink rhyolite in late magmatic activity. A very old temple is situated on the Riwasa hill. Some field photographs which were taken during field work are shown in (**Figure 3A-F**). The microgranular enclaves and mafic xenoliths are also very common features of Riwasa rhyolites. Porphyritic rhyolites display similar mineralogy of medium grained granite whereas non-porphyritic variety of rhyolite is very unique in their mineralogical assemblages. It consists of high temperature sanidine mineral and embayed quartz.

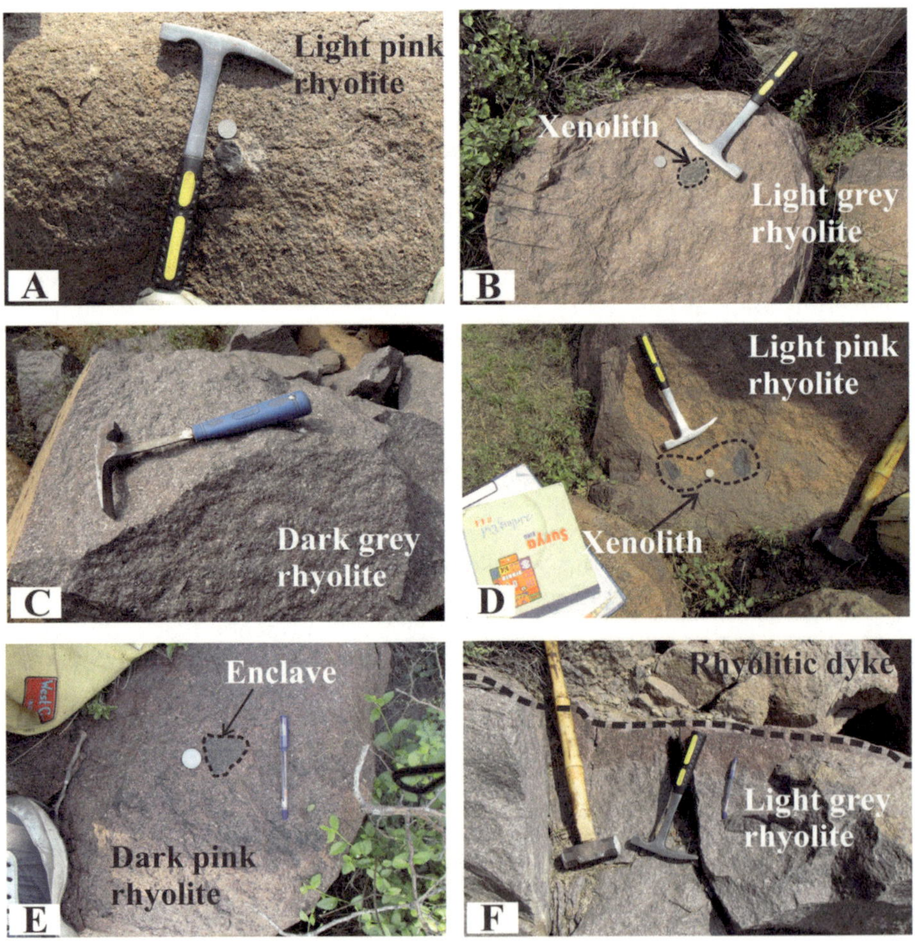

Figure 3.
Field photographs collected from Riwasa hill show (A) light pink rhyolite (B) xenolith present in light gray rhyolite (C) dark gray rhyolite (D) xenoliths present in light pink rhyolite (E) micro-granular enclave present in dark pink rhyolite (F) rhyolitic dyke cutting across light gray rhyolite.

3.2 Nigana hill

The Nigana Ring Complex (NRC) is a stock-like and ring-shaped granitic intrusion having a dimension of 2.5 × 1.5 km^2. The country rocks exposed around NRC are mainly gray granite bodies, that are intruded by a pale yellow to reddish pink and biotite granitic bodies in later stage of magmatism. Nigana granites of sub-solvus to hypersolvus nature indicate variable cooling histories on variable temperature–pressure conditions of parental magma. The granitic intrusions are of elliptical or circular shape and exhibit homogenous, massive and free from any flow structures. Post consolidation joints are very common persistent structures observed in NRC. The granites of NRC show medium to coarse grained textures. The field photographs of NRC are shown in (**Figure 4A-F**). Boulder bed, blast rock-material, dykes, high mineralized granitic surface, sharp contact between gray granite and pink granite, F- and Cl-rich biotite in biotite granite [17], sulphides mineral leaching, weathering

Figure 4.
Field photographs collected from Nigana hill show (A) boulder beds settled in Nigana granites (B) blast rock material are present along the jointed granitic surfaces (C) granitic dyke cutting across pink granite (D) high mineralized surface and dykes exposed on granite (E) dyke intrusion between gray granite and pink granite (F) sulphide minerals leaching from pink granite.

Figure 5.
Field photographs collected from Dharan hill show (A) quartz vein present in dark gray granite (B) boulder bed of granite closely packed by wind flow (C) gray granite variety (D) xenoliths present in pink granite.

Figure 6.
Field photographs collected from Dulheri hill show (A) highly jointed and fractured granitic surface (B) contact between gray and pink granite (C) pegmatite vein and quartz vein across light gray granite (D) xenolith present in light pink granite.

products, pegmatitic rim, altered feldspar surfaces and quartz veins are very common characteristics of NRC.

3.3 Dharan hill

The neighboring hill nearby Tusham is Dharan which has dimensions of 0.7 × 0.8 km. The main rock-types of this hill are granites with gray to pink color (**Figure 5A-D**). Quartz veins, xenoliths of basic composition, spheriodal weathering, quartz porphyry and boulder beds are observed in this hill-lock.

3.4 Dulheri hill

The neighboring hill nearby Nigana hill is Dharan which has dimensions of 1.1 × 0.9 km. Gray colored granite has been intruded by pink granite. It suggests that pink granite was formed in the later stage of crystallization. Pegmatitic rim and veins, iron encrustation, vertical columns, joints, fractures, sharp contact between two granites and postmagmatic alterations are the distinctive features of these litho-units. Some photographs of important physical features are taken during field work (**Figure 6A-D**).

4. Petrographical relationships and mineralogical assemblages

The photomicrographs display the rhyolitic textures in which xenoliths, sanidine, embayed and droplike quartz morphology are very common characteristics of rock-type of Riwasa hill (**Figure 7A-F**).

4.1 Gray rhyolite

Under microscope, the thin-section of gray rhyolite display porphyritic to sub-porphyritic and spherulitic textures. It includes plagioclase, quartz, sanidine and K-feldspar (minor) with biotite, chlorite, magnetite, apatite, sphene, ilmenite, rutile, monazite, Fe-Ti oxides and zircon. Sericite, epidote and kaolinite are the secondary minerals which are formed by the alteration of feldspars. Quartz phenocrysts occur as bipyramidal, drop-like, rounded, sutured and embayed in quartz due to magma resorption caused by changes in P–T conditions and may suggest a change in magma-composition around the embayed grains [18].

4.2 Pink rhyolite

At many places, pink rhyolites occur as extrusions in gray rhyolite. It shows spherulitic, granophyric, glomeroporphyritic, microcrystalline and perlitic textures with partially altered mineralogy. Essential minerals include K-feldspar, quartz, sanidine, biotite and plagioclase whereas accessory minerals are sphene, apatite, zircon, chlorite, ilmenite, rutile, monazite and magnetite. Epidote, sericite, calcite and kaolinite are the secondary minerals. K-feldspar is microperthitic and spherulitic at many places. Plagioclase phenocrysts are albite twinned. Myrmerkites texture developed at the junction of microperthite and spherulite. Some welded tuffs are directly associating with pink rhyolites, consisting of orthoclase, quartz, plagioclase and opaques and displaying a microcrystalline texture. Embayment, rounded quartz and perthite phenocrysts present in pink variety of

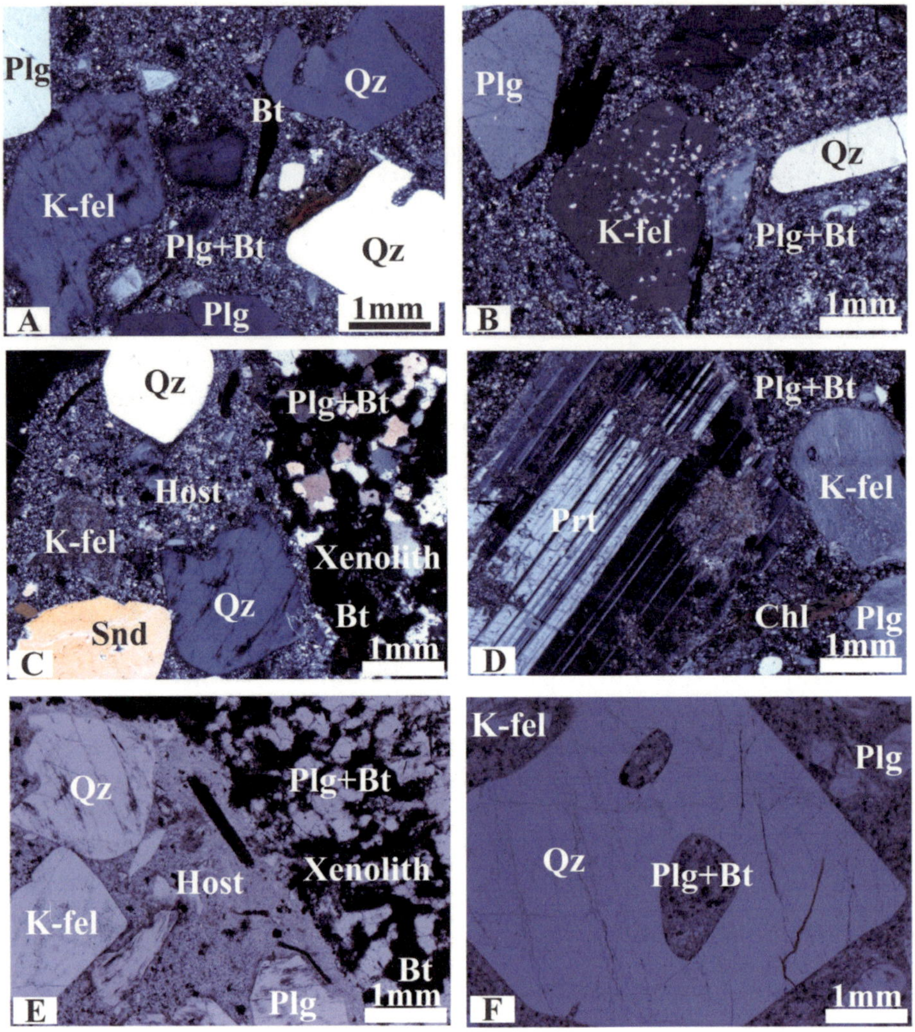

Figure 7.
(A-F) microphotographs collected from microscopic study show different textures present in different color of rhyolite of Riwasa hill.

rhyolite suggest their partial resorption prior to eruption [19]. Embayed phenocrysts may represent highly localized resorption due to convection around gas bubble, or may represent a growth phenomenon. All rock samples of rhyolite contain Fe-Ti oxide minerals and short, prismatic and fine crystals of zircon which are scattered in the groundmass. There are ubiquitous sericitization and kaoliniteitization of potash feldspar.

4.3 Tuffaceous rhyolite

It is very fine-grained variety of rhyolite exhibiting non-porphyritic texture. Quartz, plagioclase, biotite and K-felspar (minor) are essential minerals whereas zircon, apatite and ilmenite are accessory minerals. The mineral composition of this variety (non-porphyritic) is very similar to gray rhyolite. Quartz also occurs as veins that traverse the groundmass. Sanidine occurs as medium to large phenocrysts and shows Carlsbad twinning. Perthite and orthoclase occur as subhedral crystals and show vein type perthitic textures and Carlsbad twinning respectively. Further, perthite altered to sericite and kaolinite whereas sanidine and orthoclase altered to epidote. Short, prismatic and fine crystals of zircon are encountered in the groundmass. All samples contain equate opaque grains scattered in the groundmass.

Some photomicrographs (**Figure 8A-F**) represent the best granitic textures of NRC in which albite, chlorite and altered K-feldspar are very common. The granites present in Nigana, Dharan and Dulheri are of similar composition and their mineralogy is also very similar. The main rock-types of these three hills are gray granite, pink granite and biotite granite with variable size of dykes. The plagioclase feldspar is very dominant mineral in Dharan granite (**Figure 9A-F**) whereas K-feldspar mineral is dominant in Dulheri granite (**Figure 10A-F**).

4.4 Gray granite

The gray granite which is generally porphyritic and cut by numerous felsic dykes consists essentially of plagioclase feldspar (albite to andesine), quartz, K-feldspar and biotite whereas zircon, apatite, sphene, rutile, fluorite, hematite, allanite, goethite, monazite and ilmenite are accessory minerals. Chlorite and sericite are alteration

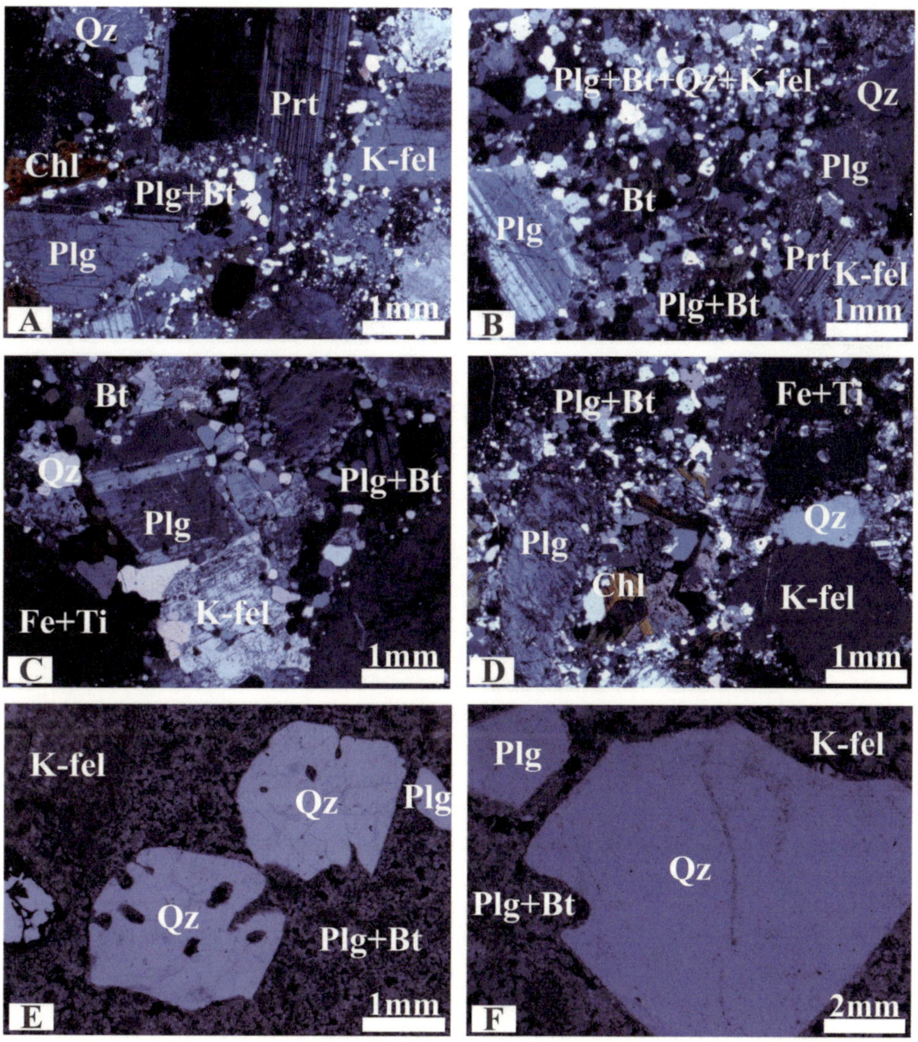

Figure 8.
(A-F) microphotographs collected from microscopic study show different textures present in different color of granites of Nigana hill.

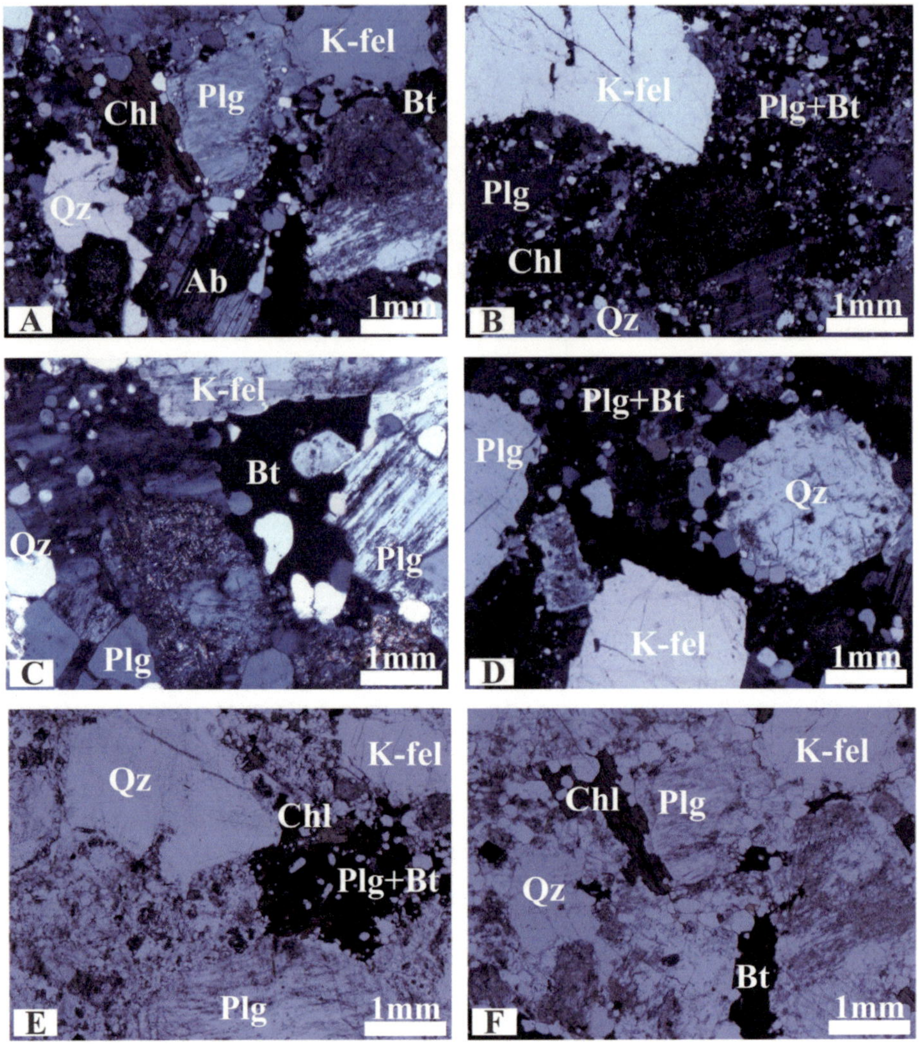

Figure 9.
(A-F) microphotographs collected from microscopic study show different textures present in different color of granites of Dharan hill.

product phases. The NRC granites exhibit porphyritic, hypidomorphic, granophyric and micrograniitic texture, in which quartz is dominant phenocryst followed by plagioclase and orthoclase. Quartz crystals are the most abundant phase in the rock with an average modal content of 35%. Quartz occurs in two different varieties; medium and fine grained. The medium subhedral shape commonly occurred as subrounded to rounded phenocrysts. Numerous poikilitic inclusions of fine grained plagioclase laths are sporadic in the quartz phenocrysts. The fine grained quartz consists of anhedral shaped constituting part of the groundmass. The dense plagioclase (albite and oligoclase) laths form the bulk of groundmass as well as poikilitic inclusions in quartz and K-feldspar. K-feldspar is represented by orthoclase as subhedral to anhedral microphenocrysts with abundant inclusions of albite laths. Among the accessory minerals which are abundant in most of the samples, magnetite, hematite, fluorite, ilmenite, allanite are the most common followed by rutile, pyrochlore, sphene, monazite, goethite and apatite. Zircon is revealed as rhombic fine-grained, subhedral to euhedral zoned crystals, accumulated in the form of cluster aggregates.

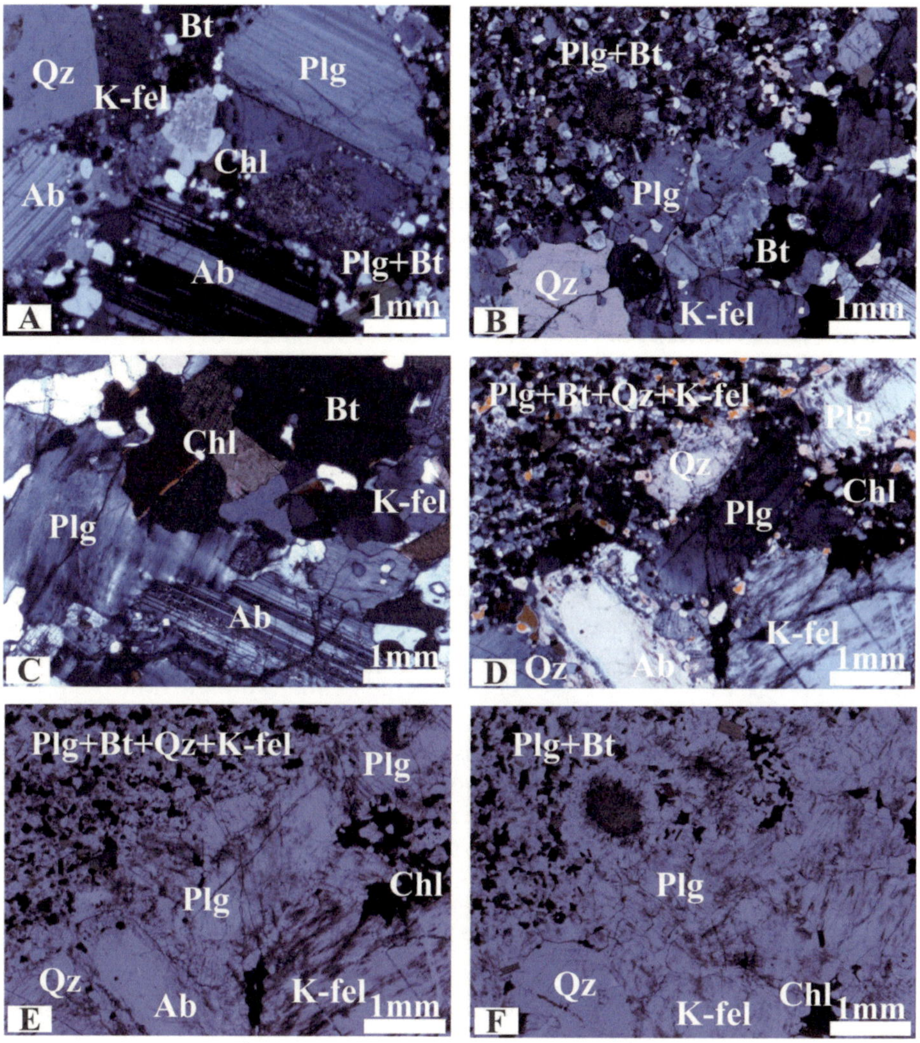

Figure 10.
(A-F) microphotographs collected from microscopic study show different textures present in different color of granites of Dulheri hill.

4.5 Pink granite

The pink granite or alkali feldspar granite consists of K-feldspar, quartz, plagioclase as essential minerals, whereas zircon, fluorite, chlorite, ilmenite, rutile, sphene, apatite, hematite, goethite, allanite, pyrochlore, thorite, doverite are accessory mineral phases. Perthites are characterized by cloudy, patchy, incoherent and extensive coarsening which are result of feldspar-fluid interaction at subsolidus temperature that leads to the replacement of albite at the margin of perthite. Albite is identified as lath-shaped crystal which exhibits polysynthetic twining. At some places, some mica flakes (mainly biotite) are also scattered along the margin of perthite as post-magmatic phase due to accumulation of residual fluid. Orthoclase is medium grained and subhedral with Carlsbad twining. Plagioclase occurs as lath-shaped crystal and showing 12^0 to 19^0 extinction angles. Biotite is strongly pleochroic (X = yellow brown; Y = reddish brown; Z = olive green), corroded and partially or completely resorbed. Along NE–SW direction, pink granites display their intrusions through the gray granite which is of high mineralization potentials. It also contains pleochroic haloes around minute zircon crystals.

4.6 Biotite granite

This variety of granite in NRC has minor exposures on the southwestern flank of the hill. It consists mainly of quartz, K-feldspar, plagioclase and biotite as essential minerals whereas zircon, apatite, hematite, chlorite, monazite, sphene, fluorite and chlorite are accessory minerals. Biotite crystals are subhedral (fine to medium grains) and they are scattered as cluster aggregates in the rock. Some biotites are altered to chlorite partially or completely. Sphene, as euhedral to subhedral crystal, is the most abundant accessory mineral. Apatite and zircon display subhedral to euhedral prismatic to acicular form. They are commonly associated with the biotite flakes and occur as scattered crystals in the rock. Quartz occurs as fine to medium grained granular aggregates filling the interstices between plagioclase and K-feldspar. On the north-western margin of the NRC and at numerous contact zones between gray and pink granites, porphyritic granite varieties (red colored granite and biotite granite) are exposed. In this zone, altered perthite, albite and quartz are essential minerals. It contains small clots of beached biotite, hematite and fluorite. This red color granite and biotite granite have similar fabric to the gray granite which is cut by the same swarms of felsic dykes and is therefore thought to be altered granite varieties that have been affected by metasomatic fluids. Biotite is scattered commonly with high contents of fluorite and chlorites indicating hydrothermal fluid activity in NW Indian shield [20]. The similar type biotite mineral with some halogens content is reported from the studied areas. It also suggests that the rock-suites of TRC might be has undergone various complex geological processes i.e. hydrothermal fluid activity, post-magmatic alteration and crustal contamination in uprising magma.

4.7 Acidic dykes

Acid dykes of granitic to rhyolitic compositions exhibit variable grain size and predominately consist of quartz, alkali feldspar, plagioclase as essential minerals with accessory minerals of magnetite, hematite, chlorite, fluorite, zircon, ilmenite, rutile, sphene, apatite and monazite. Phenocrysts of perthite are mostly altered to kaolinite and sericite at many places. Hematite is well preserved as phenocrysts in the fine grained groundmass. Zircon is present as colorless inclusions in the perthite as well as in the groundmass, displaying prismatic habit. Silver gray anhedral ilmenites resembling intergrowth with feldspar phenocrysts as well as in the groundmass. Some ilmenites are hydrothermally altered to leucoxene as minute white internal reflections. Fine grained, light yellow colored monazite is associated with quartz in the groundmass. Some opaque minerals consisting of fine grained plagioclase and biotite displays mafic composition. At some places, dykes of varied dimensions (0.4–5 meter) represent sharp contact between gray and pink varieties of rhyolites and granites in the region.

4.8 Microgranular enclaves

Microgranular enclaves are dominant component in both granitic and rhyolitic rocks and may also provide genetic linkage of the magma source, geodynamic setting and interaction between mantle and crustal melts. However, there are many contradictions between the three main genetic hypothesis that were advocated for the origin of such enclaves and xenoliths, –: including whether they are cognate cumulate, refractory or restitic fragments from granitic source rocks, and/or globules of mafic magma that have mingled or partially mixed with crustal felsic melt [21]. These physical features reported the order of phase formation during cooling of magma crystallization process and explain that xenoliths/microgranular enclaves are older phase than studied granitoids. Under microscope, enclaves possess mafic minerals especially; biotite and plagioclase in their groundmass. Now, it can be assumed that parental magma, from which studied granitoids are derived, could be of mafic nature. During the uprising magma processes, some crustal materials are partially mixed which change it to intermediate composition.

5. Petrogenetic aspects

Based on the field investigation, petro-mineralogical observations and geochemistry, it is clear that TRC is extension of MIS in southwestern Haryana. The geological features i.e. (F and Cl-rich biotite, pegmatite rim, xenoliths, microgranular enclaves, high mineralized veins, joints, fractures, vertical columns, spheroidal weathering, quartz porphyry, dykes and altered mineralogy) suggest very clear similarities with A-type, anorogenic and within-plate magmatic suites as early reported MIS in NW Indian shield. The volcano-plutonic rock associations in MIS were studied in the past by many workers [2, 4, 19, 22, 23]. TRC is assumed to be formed from three major lithological associations having (i) acid volcanic and plutonic rocks representing the first stage of igneous activities in MIS [24, 25], (ii) discordant plutons and bosses as the second stage granites of different colors and (iii) dykes of microgranites and rhyolites cutting across the host rocks are the third stage. Different types of granites are recognized as coarse to medium grained gray, grayish green granites, fine to coarse pink granites with quartz porphyry, coarse-grained porphyry and biotite granites from Khanak, Devsar, Dadam and Tusham [4]. Mineralization of porphyry copper and tin deposits was documented from rock-suites of TRC which was considered as an extension of MIS [26]. From the geological information given in the present study, the rock-types of Riwasa, Nigana, Dharan and Dulheri can be subdivided into three main categories: (i) rhyolite as volcanic phase formed during first stage of igneous activity, (ii) granites of different colors as plutonic phases formed during second stage of igneous activity and (iii) dykes of microgranular granites and rhyolites were intruded during third and last stage of magmatism. The high heat production nature and high mineralization potentiality of A-type Malani rocks are very important characteristics which can be implemented on the rock-types of TRC for mineral prospecting and exploration purposes.

6. Analytical methods

A large number of samples (16) including granite, rhyolite was collected for detail petrographical and geochemical studies. Thin sections of representative samples are studied under microscope. The petrographical study and whole-rock geochemical analysis were carried out at the Wadia Institute of Himalayan Geology (WIHG), Dehradun, India (**Table 1**). To describe the geochemical characteristics of investigating areas, representative samples from TRC were selected for geochemical analysis. Major oxides and selected trace element analysis were carried out from powder pellets methods using X-Ray Fluorescence Spectrometer. Loss-on-ignition was determined by heating a separate aliquot (0.5 gm rock powder) of each representative sample at 1000^0 C for 5 hrs. Rare earth elements (REE) of the

SAMPLE	RWNA	RWNB	RWNC	RWND	NGNA	NGNB	NGNC	NGND	DHNA	DHNB	DHNC	DHND	DUNA	DUNB	DUNC	DUND
Location		**Riwasa**				**Nigana**				**Dharan**				**Dulheri**		
SiO2	71.09	71.05	74.59	74.65	70.41	65.70	72.68	73.94	75.26	73.40	73.83	67.18	71.35	70.30	70.02	70.55
TiO2	0.17	0.16	0.19	0.20	0.42	0.59	0.21	0.19	0.06	0.12	0.14	0.57	0.29	0.34	0.32	0.34
Al2O3	12.15	12.75	13.16	12.71	13.99	14.73	13.34	12.8	13.64	14.23	13.31	14.73	14.25	14.30	15.02	15.24
Fe2O3	6.2	4.94	2.21	2.85	3.9	4.79	2.93	2.58	1.05	1.41	1.42	4.28	3.03	3.19	3.32	3.25
MnO	0.07	0.07	0.04	0.05	0.06	0.07	0.04	0.04	0.02	0.01	0.02	0.07	0.06	0.07	0.06	0.06
MgO	0.31	0.27	0.16	0.16	0.45	0.68	0.2	0.17	0.07	0.15	0.16	0.75	0.30	0.32	0.31	0.34
CaO	0.89	0.95	1.22	1.11	1.66	2.21	1.05	1.19	0.51	0.86	1.39	2.10	1.30	1.58	1.58	1.24
Na2O	2.59	2.64	2.48	2.55	2.61	2.67	2.62	2.54	3.41	3.52	3.33	2.63	2.86	2.82	2.84	2.95
K2O	5.08	5.12	5.55	5.40	5.73	5.77	5.35	5.41	5.65	5.85	5.89	5.72	5.71	5.74	5.78	6.18
P2O5	0.02	0.02	0.04	0.03	0.1	0.17	0.04	0.03	0.02	0.02	0.04	0.18	0.06	0.09	0.08	0.09
Na2O + K2O	7.67	7.76	8.03	7.95	8.34	8.44	7.97	7.95	9.06	9.37	9.22	8.35	8.57	8.56	8.62	9.13
Q	34.27	33.72	36.18	33.62	29.35	22.82	34.53	35.74	32.66	28.53	29.39	24.88	29.80	28.32	27.76	26.83
Or	30.02	30.26	32.8	31.91	33.86	34.10	31.62	31.97	33.29	34.57	34.81	33.80	33.74	33.92	34.16	36.52
Ab	21.92	22.34	20.99	21.58	22.09	22.59	22.17	21.49	28.85	29.79	28.18	22.25	24.20	23.86	24.03	24.96
An	4.28	4.58	5.79	5.31	7.58	9.85	4.95	5.71	2.40	4.14	3.97	9.24	6.06	7.25	7.32	5.56
Cor	0.82	1.19	0.95	0.72	0.72	0.48	1.43	0.67	1.04	0.59	...	0.83	1.14	0.79	1.41	1.66
Hy	0.77	0.67	0.4	0.40	1.12	1.69	0.5	0.42	0.17	0.37	...	1.87	0.75	0.80	0.77	0.85
Il	0.15	0.15	0.09	0.11	0.13	0.15	0.09	0.09	0.04	0.02	0.04	0.15	0.13	0.15	0.13	0.13
Ap	0.05	0.05	0.09	0.07	0.23	0.39	0.09	0.07	0.05	0.05	0.09	0.42	0.14	0.21	0.19	0.21
Hem	6.2	4.94	2.21	2.85	3.9	4.79	2.93	2.58	1.05	1.41	1.42	4.28	3.03	3.19	3.32	3.25
Ru	0.12	0.1	0.18	0.18	0.45	0.65	0.21	0.18	0.05	0.14	...	0.63	0.28	0.33	0.32	0.35

SAMPLE	RWNA	RWNB	RWNC	RWND	NGNA	NGNB	NGNC	NGND	DHNA	DHNB	DHNC	DHND	DUNA	DUNB	DUNC	DUND
Location		**Riwasa**				**Nigana**				**Dharan**				**Dulheri**		
Sc	7.4	6.5	4.9	5	6.9	7	4.9	5	BDL	BDL	BDL	BDL	BDL	BDL	BDL	BDL
V	9	6	9	8	24	40	9	6	4.00	4.00	6.00	38.00	12.00	18.00	16.00	19.00
Cr	23	19	17	4	16	11	14	22	7.00	6.00	7.00	17.00	8.00	8.00	8.00	8.00
Ni	18	12	2	3	4	7	2	2	—	—	—	3.00	1.00	2.00	2.00	1.00
Cu	26	17	2	9	4	8	10	2	1.00	1.00	1.00	7.00	4.00	7.00	4.00	5.00
Zn	43	47	26	46	48	72	36	31	8.00	14.00	20.00	64.00	50.00	64.00	43.00	57.00
Ga	15	17	18	19	19	19	18	18	20.00	20.00	20.00	19.00	19.00	20.00	19.00	20.00
Rb	357	345	289	346	339	324	324	329	348.00	368.00	358.00	284.00	320.00	320.00	292.00	316.00
Sr	54	52	77	69	105	173	65	64	11.00	33.00	48.00	177.00	96.00	112.00	115.00	111.00
Y	53	58	51	61	58	51	56	58	62.00	73.00	70.00	47.00	59.00	54.00	48.00	55.00
Zr	194	194	230	220	335	377	222	224	58.00	159.00	134.00	371.00	288.00	290.00	292.00	306.00
Nb	24	24	18	23	24	24	20	21	48.00	37.00	37.00	20.00	20.00	21.00	18.00	21.00
Ba	365	403	746	631	899	1100	627	620	293.00	332.00	307.00	1201.00	855.00	869.00	878.00	923.00
Pb	42	49	35	47	37	43	39	33	47.00	23.00	29.00	37.00	50.00	50.00	37.00	40.00
Th	94	103	121	101	96	65	98	105	35.00	93.00	65.00	60.00	120.00	96.00	96.00	104.00
U	7.5	9.8	13.8	9.7	6.8	6.0	9.7	8.3	15.50	35.60	26.30	5.10	9.80	7.90	7.50	8.70
Ba /Rb	1.02	1.17	2.58	1.82	2.65	3.40	1.94	1.88	0.84	0.90	0.86	4.23	2.67	2.72	3.01	2.92
Rb/Sr	6.61	6.63	3.75	5.01	3.23	1.87	4.98	5.14	31.60	11.15	7.46	1.60	3.33	2.86	2.54	2.85
Rb/Ba	0.98	0.86	0.39	0.55	0.38	0.29	0.52	0.53	1.19	1.11	1.17	0.24	0.37	0.37	0.33	0.34
Ba/Sr	6.76	7.75	9.69	9.14	8.56	6.36	9.65	9.69	26.64	10.06	6.40	6.79	8.91	7.76	7.63	8.32
Th/U	12.53	10.51	8.77	10.41	14.12	10.83	10.10	12.65	2.26	2.61	2.47	11.80	12.20	12.20	12.80	12.00

SAMPLE	RWNA	RWNB	RWNC	RWND	NGNA	NGNB	NGNC	NGND	DHNA	DHNB	DHNC	DHND	DUNA	DUNB	DUNC	DUND
Location		**Riwasa**				**Nigana**				**Dharan**				**Dulheri**		
La	263.9	262.72	279.07	264.74	283.24	264.72	238.86	272.07	48.17	86.08	55.79	180.39	224.20	186.06	247.06	200.98
Ce	528.51	528.56	535.3	529.56	555.14	529.54	457.99	533.47	90.18	168.30	107.79	336.20	417.48	344.12	439.80	374.06
Pr	58.72	59.31	56.34	59.36	59.36	59.34	48.79	56.93	10.13	18.87	12.22	37.09	45.31	37.29	46.64	41.00
Nd	189.81	189.3	182.08	190.42	189.59	190.4	156.97	182.38	33.48	64.03	40.99	126.35	145.80	123.41	151.15	136.83
Sm	30.26	30.36	27.26	31.34	29.3	31.32	24.13	25.63	7.73	13.21	9.24	21.93	24.37	20.87	23.95	24.01
Eu	1.73	1.57	2.58	1.56	1.91	1.58	1.93	1.99	0.95	0.99	0.99	3.57	2.17	2.28	2.73	2.60
Gd	28.85	29.19	25.56	28.96	27.64	28.98	22.5	23.8	8.37	11.86	9.44	20.72	22.99	19.78	22.22	22.47
Tb	3.69	3.83	3.09	3.62	3.37	3.66	2.81	2.56	1.74	1.84	1.65	2.73	2.96	2.54	2.82	2.96
Dy	18.33	19.45	14.29	19.26	15.74	19.22	13.27	11.14	11.59	10.31	10.00	13.32	14.03	11.84	12.65	13.70
Ho	3.61	3.83	2.61	3.62	3.12	3.64	2.57	1.98	2.93	2.49	2.42	2.86	3.02	2.47	2.62	2.90
Er	9.74	10.42	7.2	10.18	8.17	10.24	6.91	5.48	8.74	7.33	6.90	7.37	7.99	6.54	7.18	7.71
Tm	1.39	1.54	0.99	1.42	1.14	1.46	0.98	0.72	1.69	1.29	1.21	1.15	1.18	0.96	1.00	1.12
Yb	9.27	9.86	6.62	9.82	7.55	9.84	6.23	4.79	11.75	9.17	8.80	6.77	7.42	6.02	6.43	7.03
Lu	1.33	1.46	0.9	1.42	1.08	1.44	0.92	0.73	1.67	1.33	1.22	0.95	1.01	0.85	0.85	0.96
ΣREE	1149.14	1151.4	1143.89	1155.28	1186.35	1155.38	984.86	1123.67	239.12	397.10	268.66	761.40	919.93	765.03	967.10	838.33
ΣLREE	1071.2	1070.25	1080.05	1075.42	1116.63	1075.32	926.74	1070.48	192.90	354.09	229.49	717.33	871.53	726.23	924.13	791.48
ΣHREE	76.21	79.58	61.26	78.3	67.81	78.48	56.19	51.2	48.48	45.62	41.64	55.87	60.60	51.00	55.77	58.85

Table 1.
The whole-rock geochemical data of acid volcanic and plutonic rocks from Tusham Ring Complex, Southwestern Haryana, India.

Sample	RWN10	RWN10	RWN20	RWN20	RWN41	RWN41	NGN32B	NGN32B	NGN19	NGN19	NIN30	NIN30	NIN2	NIN2/X	DUN2	DUN2
Points ID	7 / 1.	8 / 1.	55 / 1.	57 / 1.	100 / 1.	109 / 1.	175 / 1.	177 / 1.	233 / 1.	234 / 1.	11 / 1.	14 / 1.	46 / 1.	47 / 1.	174 / 1.	194 / 1.
Oxides																
Na2O	2.96	7.08	3.33	7.66	2.71	6.21	6.05	1.39	7.05	7.82	7.71	11.22	1.15	8.30	6.71	6.37
MgO	0.00	0.01	0.00	0.00	0.00	0.00	0.00	0.00	0.00	0.00	0.11	0.00	0.01	0.00	0.00	0.00
Al2O3	18.61	24.07	18.50	24.33	17.72	26.12	26.33	17.91	24.72	24.34	23.38	18.76	17.95	23.49	25.45	26.62
P2O5	0.01	0.00	0.00	0.02	0.00	0.02	0.00	0.00	0.08	0.04	0.00	0.01	0.01	0.00	0.02	0.03
Cr2O3	0.00	0.00	0.01	0.00	0.00	0.00	0.00	0.00	0.00	0.00	0.05	0.00	0.00	0.00	0.00	0.00
MnO	0.00	0.00	0.00	0.00	0.09	0.05	0.03	0.00	0.00	0.00	0.00	0.00	0.00	0.00	0.00	0.00
CaO	0.22	6.51	0.28	6.57	0.16	8.83	9.29	0.06	5.78	6.26	4.80	8.01	0.09	5.70	7.92	8.91
K2O	11.73	0.93	10.91	0.16	12.06	0.55	0.31	14.93	0.26	0.16	0.30	0.02	14.89	0.17	0.37	0.39
TiO2	0.02	0.02	0.06	0.02	0.00	0.00	0.00	0.00	0.00	0.00	0.00	0.06	0.11	0.01	0.02	0.00
FeO	0.28	0.26	0.14	0.22	0.30	0.05	0.21	0.07	0.00	0.00	0.10	0.02	0.10	0.11	0.03	0.12
SiO2	63.27	58.35	65.19	59.19	64.38	55.98	55.03	63.46	61.04	58.69	61.71	60.94	64.86	62.17	56.61	55.26
NiO	0.03	0.00	0.00	0.00	0.02	0.00	0.00	0.01	0.08	0.01	0.00	0.00	0.04	0.00	0.00	0.00
V2O3	0.00	0.00	0.00	0.00	0.00	0.00	0.00	0.02	0.00	0.00	0.00	0.00	0.00	0.00	0.00	0.00
F	0.14	0.00	0.19	0.06	0.01	0.00	1.42	0.00	0.00	0.00	0.07	0.84	0.00	0.00	0.00	0.00
Cl	0.05	0.00	0.01	0.01	0.06	0.01	0.00	0.01	0.00	0.00	0.01	0.00	0.03	0.01	0.00	0.00
Total	97.32	97.23	98.63	98.23	97.52	97.81	98.67	97.89	99.01	97.31	98.24	99.89	99.23	99.96	97.12	97.70
Formula based on Oxygen 8 atoms																
Na	0.27	0.63	0.30	0.67	0.25	0.55	0.54	0.13	0.61	0.69	0.67	1.00	0.10	0.71	0.60	0.57
Mg	0.00	0.00	0.00	0.00	0.00	0.00	0.00	0.00	0.00	0.00	0.01	0.00	0.00	0.00	0.00	0.00
Al	1.03	1.30	1.00	1.30	0.98	1.41	1.44	1.00	1.30	1.31	1.24	1.01	0.98	1.23	1.38	1.45

Sample	RWN10	RWN10	RWN20	RWN20	RWN41	RWN41	NGN32B	NGN32B	NGN19	NGN19	NIN30	NIN30	NIN2	NIN2/X	DUN2	DUN2
P	0.00	0.00	0.00	0.00	0.00	0.00	0.00	0.00	0.00	0.00	0.00	0.00	0.00	0.00	0.00	0.00
Cr	0.00	0.00	0.00	0.00	0.00	0.00	0.00	0.00	0.00	0.00	0.00	0.00	0.00	0.00	0.00	0.00
Mn	0.00	0.00	0.00	0.00	0.00	0.00	0.00	0.00	0.00	0.00	0.00	0.00	0.00	0.00	0.00	0.00
Ca	0.01	0.32	0.01	0.32	0.01	0.43	0.46	0.00	0.28	0.31	0.23	0.39	0.00	0.27	0.39	0.44
K	0.70	0.05	0.64	0.01	0.72	0.03	0.02	0.90	0.01	0.01	0.02	0.00	0.88	0.01	0.02	0.02
Ti	0.00	0.00	0.00	0.00	0.00	0.00	0.00	0.00	0.00	0.00	0.00	0.00	0.00	0.00	0.00	0.00
Fe	0.01	0.01	0.01	0.01	0.01	0.00	0.01	0.00	0.00	0.00	0.00	0.00	0.00	0.00	0.00	0.00
Si	2.97	2.68	3.00	2.69	3.01	2.57	2.55	2.99	2.72	2.68	2.78	2.79	3.01	2.76	2.61	2.54
Ni	0.00	0.00	0.00	0.00	0.00	0.00	0.00	0.00	0.00	0.00	0.00	0.00	0.00	0.00	0.00	0.00
V	0.00	0.00	0.00	0.00	0.00	0.00	0.00	0.00	0.00	0.00	0.00	0.00	0.00	0.00	0.00	0.00
Total	5.00	5.01	4.96	5.00	4.98	5.01	5.02	5.02	4.93	5.01	4.95	5.20	4.99	4.99	5.01	5.03

Table 2.
EPMA data of feldspar mineral from Tusham Ring Complex.

DOI: http://dx.doi.org/10.5772/intechopen.95836

Sample	RWN10	RWN10	RWN41	RWN41/X	NGN32B	NGN32B	DUN2	DUN2	NGN19	NGN19/X	NIN30	NIN30	NIN2	NIN2/X	NGN32B/X	NGN32B/X
Point ID	18 / 1.	19 / 1.	112 / 1.	113 / 1.	61 / 1.	62 / 1.	159 / 1.	160 / 1.	240 / 1.	246 / 1.	24 / 1.	25 / 1.	45 / 1.	52 / 1.	166 / 1.	167 / 1.
Na2O	0.28	0.09	0.11	0.02	0.17	0.14	0.10	0.13	0.17	0.21	0.16	0.08	0.08	0.02	0.08	0.06
MgO	4.20	3.58	3.04	3.59	5.10	5.19	5.27	5.09	9.66	11.21	2.95	3.30	2.87	4.54	4.93	4.99
Al2O3	12.39	11.97	12.60	12.17	12.56	12.76	11.30	11.61	12.32	12.40	13.30	13.67	13.10	13.69	11.96	12.32
P2O5	0.00	0.00	0.02	0.06	0.02	0.00	0.04	0.06	0.02	0.04	0.12	0.04	0.05	0.08	0.10	0.06
Cr2O3	0.09	0.06	0.08	0.07	0.06	0.04	0.00	0.02	0.00	0.01	0.00	0.04	0.08	0.13	0.12	0.09
MnO	0.65	0.44	0.61	0.56	0.28	0.34	0.58	0.44	0.23	0.33	0.45	0.33	0.52	0.47	0.45	0.44
CaO	0.16	0.21	0.07	0.04	0.04	0.01	0.01	0.04	0.16	0.00	0.02	0.02	0.11	0.09	0.08	0.04
K2O	9.01	7.85	8.55	8.92	9.03	9.18	8.95	9.08	9.58	9.33	8.90	8.70	9.07	9.39	9.07	9.03
TiO2	2.74	2.35	3.48	1.83	4.17	4.22	3.42	3.04	2.77	2.38	2.59	2.55	2.63	2.68	3.97	3.96
FeO	30.28	28.58	31.11	31.29	28.61	28.78	28.30	27.92	22.20	21.72	31.47	30.19	30.53	28.04	28.35	27.89
SiO2	34.36	34.78	33.94	34.38	34.24	33.95	35.54	34.68	35.28	34.73	33.44	35.02	34.39	34.59	34.60	35.01
NiO	0.09	0.00	0.19	0.17	0.00	0.00	0.00	0.04	0.12	0.08	0.00	0.00	0.00	0.08	0.13	0.00
V2O3	0.17	0.17	0.22	0.15	0.19	0.17	0.19	0.14	0.18	0.13	0.12	0.10	0.19	0.27	0.26	0.27
F	2.16	0.96	2.25	3.08	1.87	0.99	2.59	2.04	3.68	3.95	2.09	0.80	2.75	3.47	1.67	1.75
Cl	1.14	1.13	1.04	1.01	0.50	0.51	0.93	0.89	0.51	0.47	1.41	1.31	1.19	0.82	0.49	0.51
Sum	97.71	92.19	97.32	97.36	96.84	96.27	97.23	95.21	96.87	96.96	97.01	96.13	97.55	98.36	96.26	96.42
O=F, Cl	1.50	0.80	1.50	1.90	1.10	0.70	1.60	1.30	2.10	2.20	1.50	0.80	1.80	2.10	1.00	1.10
Formula based on Oxygen22																
Na	0.09	0.03	0.04	0.01	0.05	0.04	0.03	0.04	0.05	0.07	0.05	0.03	0.03	0.01	0.03	0.02
Mg	1.03	0.91	0.75	0.90	1.24	1.26	1.29	1.27	2.32	2.69	0.73	0.81	0.71	1.11	1.20	1.21
Al	2.41	2.40	2.47	2.41	2.42	2.45	2.19	2.28	2.34	2.36	2.62	2.64	2.57	2.64	2.31	2.37

Sample	RWN10	RWN10	RWN41	RWN41/ X	NGN32B	NGN32B	DUN2	DUN2	NGN19	NGN19/ X	NIN30	NIN30	NIN2	NIN2/ X	NGN32B/ X	NGN32B/ X
P	0.00	0.00	0.00	0.01	0.00	0.00	0.01	0.01	0.00	0.01	0.02	0.01	0.01	0.01	0.01	0.01
Cr	0.01	0.01	0.01	0.01	0.01	0.01	0.00	0.00	0.00	0.00	0.00	0.01	0.01	0.02	0.02	0.01
Mn	0.09	0.06	0.09	0.08	0.04	0.05	0.08	0.06	0.03	0.05	0.06	0.05	0.07	0.06	0.06	0.06
Ca	0.03	0.04	0.01	0.01	0.01	0.00	0.00	0.01	0.03	0.00	0.00	0.00	0.02	0.02	0.01	0.01
K	1.90	1.70	1.81	1.91	1.88	1.91	1.87	1.93	1.97	1.92	1.90	1.82	1.92	1.96	1.90	1.88
Ti	0.34	0.30	0.44	0.23	0.51	0.52	0.42	0.38	0.34	0.29	0.33	0.31	0.33	0.33	0.49	0.49
Fe	4.18	4.07	4.32	4.39	3.91	3.92	3.88	3.89	3.00	2.93	4.40	4.14	4.25	3.84	3.89	3.80
Si	5.67	5.92	5.64	5.77	5.59	5.53	5.83	5.78	5.70	5.60	5.60	5.74	5.72	5.66	5.67	5.71
Ni	0.01	0.00	0.03	0.02	0.00	0.00	0.00	0.01	0.02	0.01	0.00	0.00	0.00	0.01	0.02	0.00
V	0.03	0.04	0.04	0.03	0.04	0.03	0.04	0.03	0.03	0.02	0.02	0.02	0.04	0.05	0.05	0.05
Fe++	1.05	1.02	1.09	1.10	0.98	0.99	0.98	0.98	0.75	0.74	1.11	1.04	1.07	0.96	0.98	0.96
Fe+++	2.10	2.05	2.17	2.21	1.96	1.97	1.95	1.96	1.51	1.47	2.20	2.08	2.14	1.93	1.96	1.91
Total	15.80	15.48	15.65	15.77	15.69	15.73	15.64	15.69	15.84	15.94	15.74	15.56	15.67	15.70	15.67	15.61

Table 3.
EPMA data of biotite mineral from Tusham Ring Complex.

samples were determined in the same institute by Inductively Couple Plasma-Mass Spectrometer using the open system rock digestion method. Analytical precision for major elements is well within ±2 to 3% and ± 5 to 6% for trace elements. Accuracy of rare earth elements ranges from 2 to 12% and precision varies from 1 to 8%. To study mineral chemistry of acid magmatic rocks of TRC, 9 representative thin-slides of granites (6) and rhyolites (3) were selected (**Tables 2** and **3**). The analytical work was carried out by the Electron Probe Micro Analyzer (EPMA) CAMECA SXFive instrument at DST-SERB National Facility, Department of Geology (Center of Advanced Study), Institute of Science, Banaras Hindu University. Polished thin section was coated with 20 nm thin layer of carbon for electron probe micro analyses using LEICA-EM ACE200 instrument. The CAMECA SXFive instrument was operated by SXFive Software at a voltage of 15 kV and current of 10 nA with a LaB6 source in the electron gun for the generation of an electron beam. Natural silicate mineral andardite as the internal standard used to verify positions of crystals (SP1-TAP, SP2-LiF, SP3-LPET, SP4-LTAP and SP5-PET) with respect to corresponding wavelenght dispersive (WD) spectrometers (SP#) in CAMECA SX-Five instrument. The following X-ray lines were used in the analyses: F-Kα, Na-Kα, Mg-Kα, Al-Kα, Si-Kα, P-Kα, Cl-Kα, K-Kα, Ca-Kα, Ti-Kα, Cr-Kα, Mn-Kα and Fe-Kα. Natural mineral standards: flourite, halite, apatite, periclase, corundum, wollastonite, orthoclase, rutile, chromite, rhodonite and hematite standard supplied by CAMECA-AMETEK used for routine calibration, X-ray elemental mapping and quantification. Routine calibration, acquisition, quantification and data processing were carried out using SxSAB version 6.1 and SX-Results software of CAMECA. The precision of the analysis is better than 1% for major element oxides and 5% for trace elements from the repeated analysis of standards. The analytical detials are also mentioned in Sharma and Kumar [21], Sharma et al. [4], Kumar et al. [2].

7. Mineral chemistry and bulk geochemistry

The whole-rock geochemical data of major and minor oxides with calculated CIPW norms, trace elements and rare earth elements for the acid volcano-plutonic rocks, are carried out to justify our mineralogical and petrographical results. They are high in SiO_2, K_2O + Na_2O, Al_2O_3, Rb, Zr, Ba, Y, Nb, Th, U, REEs (except Eu) and low in CaO, TiO_2, MgO, V, Ni, Cr, Sr., Ti, P, Eu; typically A-type affinity. Based on their major oxide geochemistry, they were classified into two major groups i.e. rhyolite and granite (**Figure 11**). Based on the mineral chemical databank, it was investigated that K-feldspar, plagioclase and biotite are important silicate minerals in rock-formation (**Figure 12**).

8. Geodynamics of related petrologs

On the basis of worldwide data, several petrogenetic models have been proposed for the origin of A-type granitoids, including: 1) fractional crystallization of basaltic magma [27, 28]; 2) partial melting of lower crustal rocks caused by fluxing of mantle-derived fluids/melts [29]; 3) melting of a tonalitic I-type granite [30, 31], and 4) assimilation and/or magma mixing between the mafic magma and crustal melts [32, 33]. Overall mechanism related to MIS magmatic system, it was suggested that crustal-mantle interaction is the main dominant cause in the generation of anorogenic magmatism in NW, Indian shield.

There are mainly two privileges and accepted models for Malani geodynamic system: (a) Plume related extensional model [2, 3, 5, 9, 10, 15, 34] and (b)

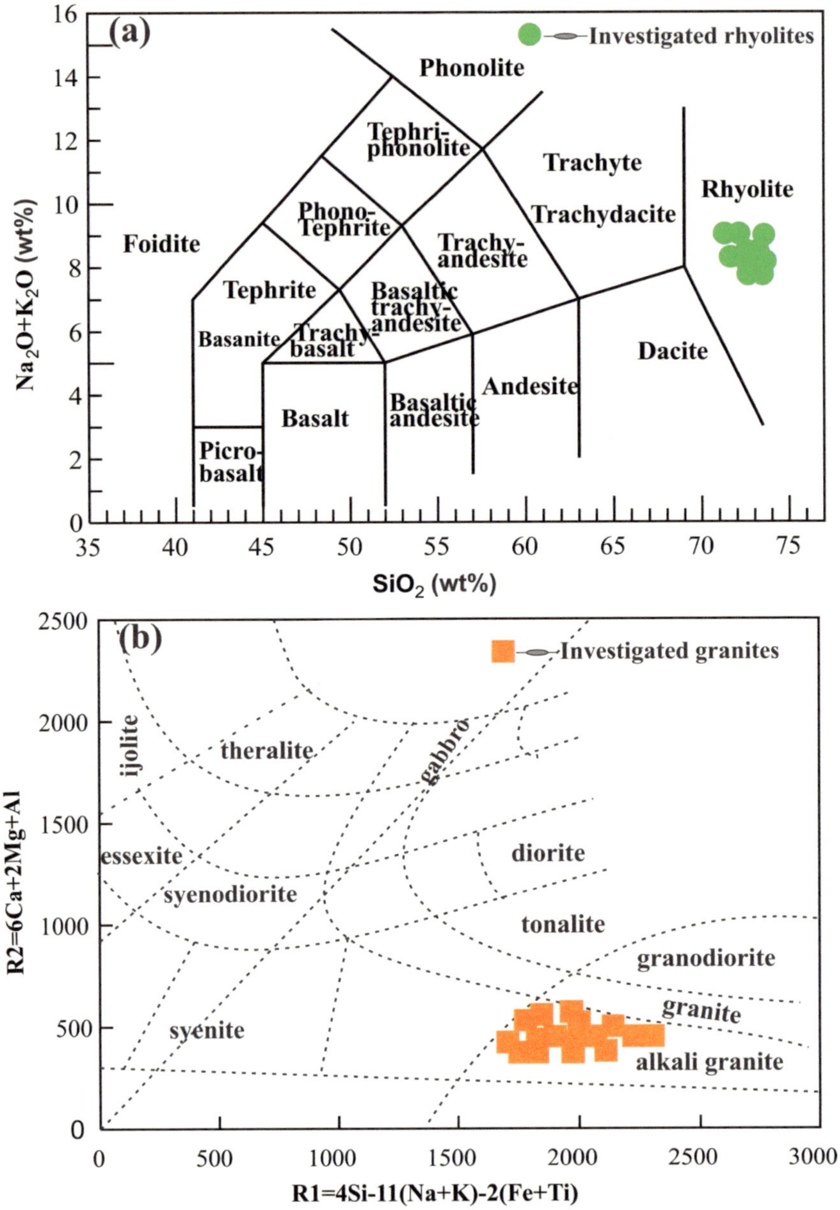

Figure 11.
(a) SiO_2 vs. Na_2O+K_2O (wt %) volcanic rocks classification binary diagram, showing that the volcanic rock samples belong to rhyolite. (b) R_1-R_2 alkaline-subalkaline compositional discrimination diagram for plutonic rocks classification. All investigating samples straddle in the alkali-granite to granite field.

Subduction model [11, 35, 36]. The present contribution is argued with plume related extensional environment. Ring structures and the cauldron subsidence are strong evidences for hot-spot magmatism in TRC and MIS respectively. The isotopic data interpreted by some previous workers, also recorded that MIS magmatism was contemporaneous with breakup of Rodinia and Pan-African thermo-tectonic event. The period ca. 732 ± 41 Ma B.P. marked a major Pan-African thermo-tectonic event of widespread magmatism of alkali granites and co-magmatic acid volcanic (anorogenic, A-type) in the TAB of the Indian Shield, Central Iran, Somalia, Nubian-Arabian Shield, Madagascar and South China [2, 5].

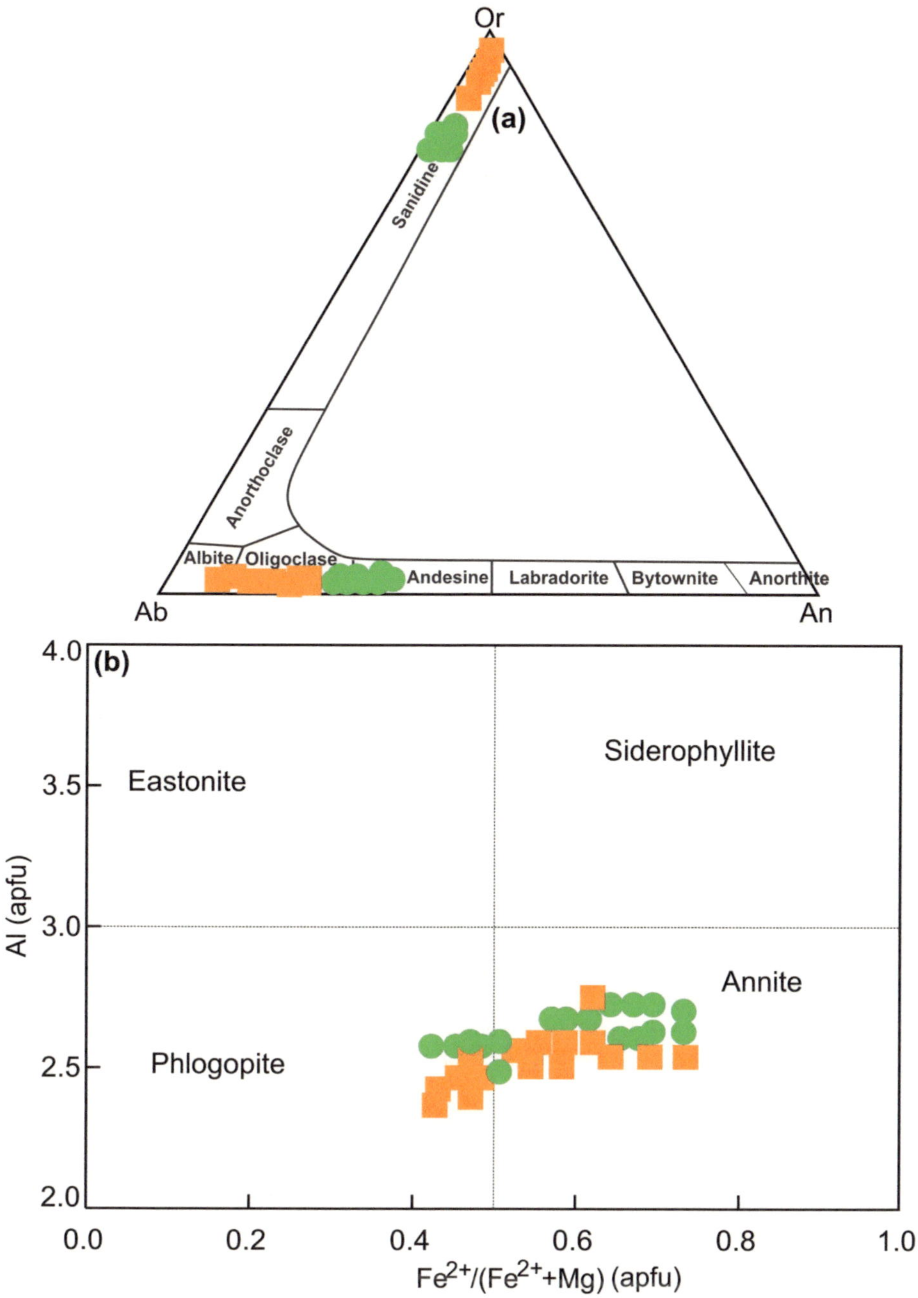

Figure 12.
(a) CIPW norm calculation of mineral chemistry data in Albite-Orthoclase-Anorthite triangular diagram explain that most of the rock samples of rhyolite and granite consist of sanidine and orthoclase as K-feldspar, respectively whereas albite, oligoclase, and andesine as plagioclase. (b) Analyzed biotite compositions in (Fe^{+2}/[Fe^{+2}+Mg]) (apfu) versus Al (apfu) classification diagram show that the biotite present in all the granitic and rhyolitic samples is annite and phlogopite.

Keeping in view, all the geological observations, it is proposed that all these micro-continents were characterized by common crustal stress pattern, rifting, thermal regime, strutian, glaciations and subsequent desiccation and similar palaeo-latitudinal positions which could be attributed to the existence of a supercontinent; "The Greater Malani Supercontinent" (reconstruction of Rodinia) [5, 12] . They were united in specific continental framework during Neoproterozoic time (Rodinia) then drifted due to some tectonic movements [11]. This assembly and subsequent breakup marked rift to drift tectonic environment which might be possible reason for the formation of new supercontinent from pre-exited parental continental supercontinent 'Rodinia' (Reconfiguration of Rodinia in new geological aspect). This complex geological setting is still a plausible concept and the present paper attests that NW India was part of Rodinia supercontinent at 780 Ma ago. To date, no detailed information about halogens role in the evolution of malani magmatism has been carried out. Our results in biotites from the investigated granitoids as well as physio-chemical features support the model, which fluorine-rich A-type granitoids may be derived from partially molten igneous rocks of tonalitic to granodiorite composition. Further investigation and experimental works are needed to better constrain and quantify the distribution of halogens in all over the TRC and MIS. In future, such re-equilibration effects of halogens are expected to carry out which will depend on the factors like cooling rate of magma and intensity of hydrothermal overprint in TAB of NW Indian shield.

9. Conclusion remarks

Based on the field information, petro-mineralogical observation and geochemistry, the TRC granitoids under study have reached on the following conclusion:

1. The rock-types exposed in Riwasa, Nigana, Dharan and Dulheri are divided into three main lithological divisions, i.e. rhyolite as first phase, granites of different colors as second phase and dykes of fine-grained granites and rhyolites as third and last phase of magmatism.

2. Based on petrographical observations, it is suggested that rhyolites show porphyritic, granophyric, glomeroporphyritic, aphyritic, spherulitic and perlitic textures whereas granites show hypidomorphic, granophyric and micrograno-phyric textures. These textures have close similarities with A-type, anorogenic and within-plate granitoids as early reported MIS rock-types behave.

3. The volcano-plutonic rock-associations and physio-chemical features indicated that the rock-types of Tusham Ring Complex have been formed throughout complex geological processes.

4. Magmatic evolution, phase petrology and geodynamic emplacement pointed out that the studied areas belonging to MIS extension in NW Indian shield might be formed under plume-related hot spot extension model.

5. Some important physical features i.e. high mineralized granitic surfaces, high mineralized veins, pegmatitic rims, iron encrustation and altered mineralogy indicate that rock-types of TRC have high mineralization potentiality which can be explored in future.

6. Based on mineral chemistry and bulk rock geochemistry, it is concluded that feldspar and biotite are important rock-forming minerals in acid volcano-plutonic rocks of TRC. Our new results also suggest that the investigating granitoids must be studied in near future to reconstruct the palaeo-existed supercontinent tectonic environment also.

Acknowledgements

The authors wish to express their thanks to Chairman, Department of Geology, Kurukshetra University, Kurukshetra, India and Director, Wadia Institute of Himalayan Geology, Dehradun, India for their support. Dr. N. V. Chalapathi and Dr. Dinesh Pandit (Faculty of Geology Department, BHU, India) are highly acknowledged for their help during EPMA analysis. The first author also expressed his thanks to the local people of Tusham area for his help during field works.

Author details

Naveen Kumar* and Naresh Kumar
Department of Geology, Kurukshetra University, Kurukshetra, India

*Address all correspondence to: naveenphdkuk@gmail.com

References

[1] Kochhar, N., 2015, The Malani Supercontinent. In K.L. Shrivastava and P.K. Srivastava (Eds.). The Frontiers of Earth Science, Chapter 6, Scientific Publishers (India), 122-136.

[2] Kumar, N., Kumar, N. and Singh, A. K., 2019, Petrology and Geochemistry of Acid Volcano-Plutonic Rocks from Riwasa and Nigana Areas of Neoproterozoic Malani Igneous Suite, Northwestern Peninsular India: An Understanding Approach to Magmatic Evolution. Geochem. Inter., 57, 645-667.

[3] Sharma, R., and Kumar, N., 2017, Petrology and Geochemistry of A-type granites from Khanak and Devsar Areas of Bhiwani District, Southwestern Haryana. Journal of the geological Society of India, 90, 138-146.

[4] Sharma, R., Kumar, N. and Kumar N., 2019, Signatures of high heat production and mineralization associated with plutonic and volcanic acidic rocks from Tosham Ring Complex, Southwestern Haryana, India. Himalayan Geology, 40 (2), 239-247.

[5] Kochhar, N., 2015, The Malani Supercontinent. In K.L. Shrivastava and P.K. Srivastava (Eds.). The Frontiers of Earth Science, Chapter 6, Scientific Publishers (India), 122-136.

[6] A. V. Grebennikov, A-type granites and related rocks: petrogenesis and classification. Russ. Geol. Geophy. 55, 1074-1086 (2014).

[7] Bonin, B., 2007, A-type granites and related rocks: Evolution of a concept, problems and prospects. Lithos, 97, 1-29.

[8] Haapala, I. and Ramo, O.T., Rapakivi granites and related rocks: An introduction, Precambrian Research, vol., 95, pp. 1-7.

[9] Kumar, N., and Vallinayagam, G., 2014, Tectonic significance of Neoproterozoic magmatism of Nakora area, Malani Igneous Suite, District Barmer, Western Rajasthan, India. Geotectonics by Springer-Verlag, Germany, 48, 239-253.

[10] Singh, A.K., Singh, R.K.B., and Vallinayagam, G., 2006, Anorogenic acid volcanic rocks in the Kundal area of the Malani Igneous Suite, Northwestern India: Geochemical and petrogenetic studies. Journal of Asian Earth Sciences, 27, 544-557.

[11] Wang, W., Cawood, P.A., Zhou, M-F., Pandit, M.K., Xiao-Ping, X., and Zhao, J-H., 2017, Low-δ18O Rhyolites From the Malani Igneous Suite: A Positive Test for South China and NW India Linkage in Rodinia. Geophysical Research Letters, 44, 10,298-10,305.

[12] Kumar N, Kumar N, Sharma R and Singh A K 2020 Petrogenesis and tectonic significance of Neoproterozoic magmatism of Tusham Ring Complex, NW Indian shield: Insight into tectonic evolution of Malani Igneous Suite and Rodinia Supercontinent; Geotectonics, **54** 428-453.

[13] A. R. Crawford and W. Compston, "The age of the Vindhyan system of peninsular India," Quart. Jour. Geol. Soc. London. 125, 351-370 (1970).

[14] D. Huang et al., "Neoproterozoic Low-δ18O Zircons Revisited: Implications for Rodinia Configuration," American Geophysical Union. 1-11 (2019).

[15] S. Dhar et al., "Sr, Pb and Nd isotope studies and their bearing on the petrogenesis of the Jalor and Siwana complexes, Rajasthan, India, Jour. Geol. Soc. India. 48 (2), 151-160 (1996).

[16] T. H. Torsvik et al., "Rodinia refined or obscured: palaeomagnetism of the Malani igneous suite (NW India)," Precambrian Res. 108, 319-333 (2001).

[17] Kumar, N., Sharma, R., Kumar, N. et al., A review on petrology and geochemistry of the Neoproterozoic Malani Igneous Suite and related rocks (Northwestern peninsular India), Petrology, 2020b, vol. 28, pp. 591-657.

[18] R. K. Gupta, "Embayed quartz crystals in Shyok volcanics, Ladakh," Contemporary Geoscientific Researches in Himalaya. 2, 65-68 (1983).

[19] A. K. Singh and G. Vallinayagam, "Radioactive element distribution and rare-metal mineralization in anorogenic acid volcano-plutonic rocks of the Neoproterozoic Malani Felsic Province, Western Peninsular India," Jour. Geol. Soc. India. 73, 837-853 (2009).

[20] P. Kaur et al., "Chlorine-rich amphibole and biotite in the A-type granites, Rajasthan, NW India: Potential indicators of subsolidus fluid-rock interaction and metallogeny," Geological Journal. 1-1 (2018).

[21] S. Kumar et al., "Geochemistry and U–Pb SHRIMP zircon geochronology of microgranular enclaves and host granitoids from the South Khasi Hills of the Meghalaya Plateau, NE India: evidence of synchronous mafic–felsic magma mixing–fractionation and diffusion in a post-collision tectonic environment during the Pan-African orogenic cycle," Geological Society, London. 457 (2017).

[22] L. S. Singh and G. Vallinayagam, "Geochemistry and petrogenesis of acid volcano-plutonic rocks of the Siner area, Siwana Ring Complex, Northwestern Peninsular India, Jour. Geol. Soc. India. 82 (1), 67-79 (2012).

[23] Vallinayagam, G., Singh, L.G., 2011. Radioactive Heat Producing felsic to intermediate Volcano–plutonic rocks of Dhiran Area, Malani Igneous Suite, Western India. Open access E-Journal Earth Science India, 4 (II), 68-97.

[24] Bhushan, S.K., 1985. Malani volcanism in Western Rajasthan. Indian Journal of Earth Sciences, 12, 58-71.

[25] Bhushan, S.K., Chittora, V.K., 1999. Late Proterozoic bimodal assemblage of Siwana subsidence structure, Western Rajasthan, India. Journal of Geological Society of India, 53, 433-453.

[26] Kochhar, N., Malani Igneous Suite: Porphyry copper and tin deposits from the Tusham Ring Complex, North Peninsular India, Geologica Carpathica, 1985, vol. 36, pp. 245-255.

[27] Brown, G.C. (1982) Calc-alkaline intrusive rocks: their diversity, evolution and relation to volcanic arcs. In: Thorpe, R.S. (Ed.), Andesites: Orogenic Andesites and Related Rocks. Wiley, Chichester, pp. 437-461.

[28] Turner, S.P., Foden, J.D. and Morrison, R.S. (1992) Derivation of some A-type magmas by fractionation of basaltic magma: an example from the Padthaway Ridge, South Australia. Lithos., v. 28, pp. 151-179.

[29] Bailey, D.K. (1978) Continental Rifting and Mantle Degassing. In: Neumann ER., Ramberg I.B. (eds) Petrology and Geochemistry of Continental Rifts. NATO Advanced Study Institutes Series (Series C-Mathematical and Physical Sciences), vol 36. Springer, Dordrecht, pp. 1-13.

[30] Creaser, R.A., Price, R.C. and Wormald, R.J. (1991) A-type granites revisited: Assessment of a residual source model. Geology, v.19, pp.163-166.

[31] Skjerlie, K.P. and Johnston, A.D. (1993) Fluid-absent melting behaviour of F-rich tonalitic genesis at mid-crustal pressures: implications for the

generation of anorogenic granites. J. Petrol. v. 34, pp. 785-815.

[32] Barboni, M. and Bussy, F. (2013) Petrogenesis of magmatic albite granites associated to cogenetic A type granites: Na-rich residual melt extraction from a partially crystallized A-type granite mush. Lithos, v. 177, pp. 328-351.

[33] Bowden, P., Black, R., Martin, R.F., Ike, E.C., Kinnaird, J.A. and Batchelor, R.A. (1987) Niger- Nigerian alkaline ring complexes: a classic example of African Phanerozoic anorogenic mid-plate magmatism. Geol. Soc. London, vol. 30, pp. 357-379.

[34] Eby, G.N., and Kochhar, N., 1990, Geochemistry and petrogenesis of the Malani Igneous Suite, North Peninsular India. Journal of the Geological Society of India, 36, 109-130.

[35] Ashwal, L. D., Solanki, A. M., Pandit, M. K., Corfu, F., Hendriks, B. W. H., Burke, K., and Torsvik, T. H., 2013, Geochronology and geochemistry of Neoproterozoic Mt. Abu granitoids, NW India: Regional correlation and implications for Rodinia paleogeography. Precambrian Research, 236, 265-281.

[36] Gregory, L.C., Meert, J.G., Bingen, B., Pandit, M. K., and Torsvik, T.H., 2009, Paleomagnetism and geochronology of the Malani Igneous Suite, Northwest India: Implications for the configuration of Rodinia and the assembly of Gondwana. Precambrian Research, 170, 13-26.